Uncle J
BATHROOM READER®

BATHROOM READERS' PRESS
ASHLAND, OREGON

"No amount of skillful invention can replace the essential element of imagination."

–Edward Hopper

"ShamWow holds 20 times its weight in liquid!"

–Vince, the ShamWow guy

UNCLE JOHN'S BATHROOM READER®

WEIRD INVENTIONS

For information, write: The Bathroom Readers' Institute,
P.O. Box 1117, Ashland, OR 97520
www.bathroomreader.com

Cover and interior design by Andy Taray / Ohioboy.com

ISBN-10: 1-60710-781-3 / ISBN-13: 978-1-60710-781-1

Library of Congress Cataloging-in-Publication Data

Uncle John's bathroom reader weird inventions.

pages cm

ISBN 978-1-60710-781-1 (pbk.)

1. Inventions--Miscellanea. 2. Inventions--Humor. 3. Curiosities and wonders. 4. American wit and humor. I. Bathroom Readers' Institute (Ashland, Or.) II. Title: Weird inventions.

T212.U65 2013

600--dc23

2012051014

Printed in the United States of America
First Printing: June 2013
1 2 3 4 5 17 16 15 14 13

THANK YOU!

The Bathroom Readers' Institute sincerely thanks the people whose advice and assistance made this book possible.

Gordon Javna
Brian Boone
Andy Taray
Christy Taray
Trina Janssen
Claudia Bauer
Dan Mansfield
Jill Bellrose
Megan Todd
Brandon Hartley
Will Harris
Jack Feerick
Chris Holmes
Erica Richman
Jeff Giles
Jon Cummings
Dave Lifton
Donny Rodriguez
Anna Zajac
Adam Bolivar
Eleanor Pierce
Eric Dodson
David Hoye
Blake Mitchum
Jennifer Frederick
Sydney Stanley
Lilian Nordland
Aaron Guzman

CONTENTS

WEIRD SCIENCE!

They say that everything that could possibly be invented has already been invented. After all, we've got flat-screen TVs that hang on the wall, cars that run on electricity, phones that hold our entire music collections, and, most impressively, Hot Pockets. But it's not true—there's always more labor to be saved or human problems to solve which will inspire some enterprising amateur Edison out there to slave over a pile of wires and molded plastic in the basement to make a gadget that forever changes the way we live.

And then there's this stuff. Welcome to *Uncle John's Weird Inventions*. The honest-to-goodness real, and really weird, gizmos in this book are all things you never knew needed to exist...because they probably don't.

In *Weird Inventions*, you'll read about:

- Machines to wash your dog, and translate its barks
- Spray-on WiFi and skin-in-a-can
- The device that makes artificial egg yolks, then perfectly slices them.

And lots more goofy gadgets, silly science, and crazy contraptions. It's patently absurd!

—Uncle John and the Bathroom Readers' Institute

BREATH-TO-ENERGY CONVERSION MASK

Trying to live your life in an environmentally-friendly fashion? Good for you, but...well, if you *really* loved your planet, you'd be doing something to help it 24/7.

Sound impossible? No, just unfeasible. Still, you'll be making major headway if you wear the Breath Charging AIRE Mask. Just slip this thing over your face, and if you can get past the fact that it makes you look like Bane from *The Dark Knight Rises*, you'll be pleased to discover that the mere act of breathing in and out activates wee wind turbines within the mask. The end result: You're creating energy that, with the appropriate attachment, can be used to charge your iPod, your iPhone...pretty much iAnything.

The AIRE mask is the brainchild of Brazilian inventor Joao Paulo Lammoglia, who trumpeted his creation in an interview with the *Daily Mail*, crowing, "It can be used indoors or outdoors, while you're sleeping, walking, running, or even reading a book." Lammoglia also added that "its energy is available 24 hours a day, seven days a week," which—what a coincidence!—is exactly how often you should be helping out the planet anyway.

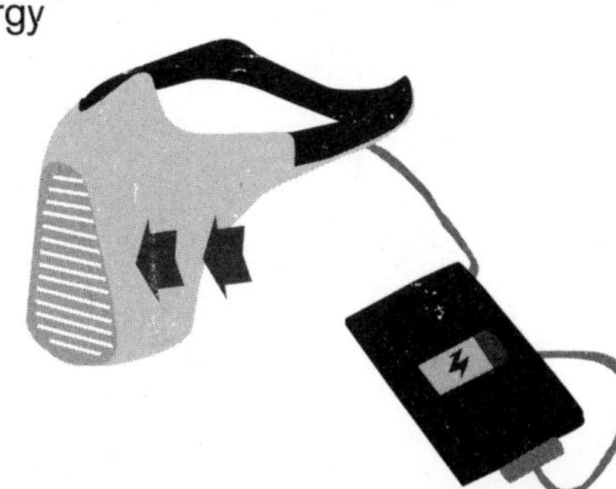

INSTANT DRUNKENNESS-REVERSING PILLS

Imagine being able to get rip-roaring drunk, raise hell for a few hours, and then pop a few pills and sober up quicker than you can say, "I'll be glad to walk on that line, officer!" Nothing bad could possibly come of that, right?

Well we're soon going to find out, because scientists appear to have unlocked the key to countering the intoxicating effects of alcohol: enzymes. A team of researchers led by Yunfeng Lu, a UCLA professor of chemical and biomolecular engineering, and Cheng Ji, a professor of biochemical and molecular biology at USC, has devised a way to package enzymes inside a nanoscale polymer shell. In non-egghead speak, they found a way to put chemicals inside your body that can change what your body does. Anyway, they tested these tiny capsules on drunk mice and found that the enzymes caused their blood alcohol levels to drop quickly and significantly.

Professor Lu stated that down the road he can envision an alcohol prophylactic or an antidote that could be taken orally. The impact of this is obvious: Without an alcohol buzz, we now have zero reasons to drink domestic beer.

INSOMNIA HELMET

The concept of the head massager–a series of thin, rubber-tipped metal "fingers" that bend to fit over one's skull and massage the surrounding area in the process–has been successfully mass-produced to the tune of being available for less than a buck. Those who wish to avoid being labeled a cheapskate may wish to upgrade to a device that is far costlier and much more elaborate, if not necessarily more effective.

Patented in 1992, the so-called "Insomnia Helmet" is lined with a series of rubber "fingers" attached to a belt which, courtesy of a small electric motor, repeatedly rotate within the helmet, soothingly massaging the head of its wearer in front-to-back fashion so that they can sleep better. The inventor's pitch for the apparatus cited the importance of head massage in civilizations both human and inhuman, referencing "the baboons who spend much time grooming one another's heads to remove insects and dirt."

Since he also noted how "the related act of head patting indicates approval or affection to children and adults both," here's hoping that someone at the U.S. Patent Office rewarded him with a few good pats for his efforts.

ARTIFICIAL CHEWING HEAD

The American junk-food consumer can always rest assured that his tasty treats have been product-tested down to the last detail in focus groups, in grocery-store trials, and, of course, in the laboratory kitchen, the birthplace of so many delicious goodies that are abominations against nature. There, our brightest culinary minds examine every factor that will determine a snack's market potential. Flavor is key, but there's also texture, aftertaste, dunk-ability, and the all-important "behavior in the mouth." Is the cracker crispy enough? Precisely how much chewing is required per bite of this soft-baked cookie? Could that new chip be somehow louder?

Clearly, snack-food impresarios spare no expense to please our palates. Yet every field of endeavor eventually bumps up against budgetary limitations, and at some point, these geniuses realized that details like crunch-testing become rather expensive when performed by actual human crunchers. So somebody devised an artificial chewing head specifically for testing snack-food crispiness. It offered a high-tech replication of mastication, complete with a built-in microphone for picking up the full spectrum of crunch noises. So nosh with confidence. Because if that chip was snacktastic enough for an animatronic head, it's snacktacular enough for you.

ARTIFICIAL LEAVES

We all know that leaves turn sunlight into energy through the process of photosynthesis, or so we've been told. In an effort to one-up nature, a team of scientists at MIT has invented an artificial leaf that works just like the real thing, only better. The science is fascinating; when placed in water and exposed to sunlight, the leaf splits the H2O molecule into its component parts of oxygen and hydrogen, producing clean, renewable fuel. (It also makes real leaves look like a bunch of freeloading jerks.)

It sounds, and is, weird, but theoretically, one artificial leaf placed in a bucket of water could provide enough energy to meet the daily electrical needs of an average home. The inventor, Daniel Nocera, has humanitarian goals in mind: Impoverished families in third-world countries could greatly increase their standard of living if they could essentially make free energy out of rainwater.

Unfortunately, development of the leaf is at a standstill due to the expense of the specialized silicon required to make it. There are cheaper ways to make energy, so Nocera's project has run out of investors. And because the leaf is a wonderful concept that could reduce pollution and improve innumerable lives, the governments of the world probably aren't going to devote precious tax dollars to help make it a reality.

FLOATING SHADE

Humanity has long struggled to invent a device that wards off the sun's harsh glare in a way that is easily portable. Well, not really—a parasol (which means "for the sun" in French) does the trick, as does an umbrella. Or a large-brimmed hat. Or a shelter of some sort.

But if you're the kind of person who still proudly uses a parasol when cavorting in the park or down the promenade, then you're probably also the kind of person who is so dainty that they tire of holding an object as light as a parasol upward for more than a few minutes.

The Floating Shade (patented in 1991) solves your problem. It's a wide, helium-filled balloon made of extra-thick, virtually unpoppable rubber that floats directly over you and your immediate personal space. It's tethered to your body with an elaborate series of ropes that strap on under your shoulders, resembling a parachute harness. It leaves both hands free to carry objects, read a book, talk on the phone, or use a walking stick. There's also an extra-large, "family-size" model for shading a group of people and their vicinity, so long as they walk very close together.

THE ELECTRIC HICCUP CURE

Everyone has a favorite home remedy for the hiccups: quickly downing a glass of water, holding your breath for a full minute, or repeatedly blowing into a paper bag, or loudly popping that same paper bag to scare the bejeezus out of the sufferer, for example. For one inventor, though, none of these old-wives'-tale cures was nearly good enough. Perhaps he objected because they don't actually attack the root cause of hiccups, which is an involuntary spasm of the diaphragm, resulting in a breath that's interrupted by the involuntary closing of the glottis. Or maybe it was because those old cures didn't involve electricity.

So he invented a cup-like appliance fitted with electrodes that make contact with the user's mouth and temple. When the user fills the cup with water and places it over his mouth to drink, the water's movement creates an electric charge that runs through the electrodes and stimulates a pair of nerves that help regulate the diaphragm. If it works, voilà! No more hiccups. And if it doesn't, perhaps the very concept of the device—a combination of low-grade waterboarding and electrocution—will scare the heck out of you and cure your hiccups anyway.

3-D PRINTABLE HAMBURGERS

Meat your future. In 2012 scientists at Maastricht University in the Netherlands managed to grow a small strip of hamburger in a lab. They even fried it up and ate it. Their research was a major step toward creating synthetic meat and even burger patties that can be made in three-dimensional bio-printers.

These 3-D printers are nothing new (one appeared in the 2001 film *Jurassic Park III*). As of 2013, however, they're only capable of creating inedible objects like the conceptual models used by architects. They can't make you something to eat, unless your idea of a tasty snack is a miniature strip mall or a plastic condominium tower.

Using the methods employed by those Dutch scientists, creating an entire lab-grown burger patty would set you back an estimated $300,000. Fortunately, a U.S.-based company called Modern Meadow is currently developing a printer capable of cranking out much cheaper artificial meat. The "bio" versions of these machines are still in their infancy. One research group recently built a bio-printer capable of producing chocolate, but more sophisticated material like meat is trickier to pull off. The folks at Modern Meadow hope to create a printer that uses stem cells much in the same way a conventional office printer uses ink. Stem cells can replicate themselves many times over and turn into more sophisticated cells. Hypothetically speaking, a bio-printer could churn out a burger, so long as it has the right material in its "bio-ink cartridge."

TOILET SNORKEL

Fun fact: If you're ever trapped in a burning building, it won't be the flames or falling debris that will kill you–it will be the smoke. A raging blaze needs oxygen to burn, so it takes that from its immediate surroundings, rudely not considering that you need that oxygen to live. Result: You die.

In 1982 William Holmes patented the Fresh-Air Breathing Device…but we're going to go ahead and call it the Toilet Snorkel, because that's what it is. Holmes realized that a good source of fresh air during fires in high-rise office buildings is deep inside the toilet, in the sewer line's vent pipe. Simply snake Holmes's slender breathing tube down through any toilet, then into the water trap and beyond, and breathe safely. Why you're wearing a charcoal-filter fitted mask connected to a tube that's buried in the toilet will be perfectly explainable to the firemen when they arrive.

BREATH DETECTOR

Dealing with *halitosis*, or bad breath, is difficult—obviously, if you knew you had bad breath, you would do something to nip it in the bud, but you can't, because it's pretty darn impossible to smell your own breath. That "blowing into your hand and trying to smell it really fast" trick just doesn't work, so instead you hope you can treat it with prevention, like brushing your teeth eight times a day or eating box after box of breath mints.

In 1925 an inventor named George Starr White was on to something, even if he just mechanized the blowing-into-your-hand trick. His Breath Detector sought to tell a person they had bad breath before someone else could. The gadget was a handheld bellows with a straw-size opening at one end of the barrel and a narrow opening at the other. A person suspecting themselves of having bad breath was to inflate the bellows by blowing into the straw side several times. Then they would deflate the bellows by placing the narrow end in their nose and squeezing. The nose-straw makes sure that you, and only you, smell the questionable breath in question.

ANTI-STICK SPRAY

To watch the video demonstration of the chemical coating called Ultra-Ever Dry on YouTube is to be amazed at the way it repels water, refined oil, wet concrete, and other liquids from seemingly any substance. But to try and explain to Joe Average the phenomenon behind why it works without causing the words "System Overload Imminent" to flash before their eyes...well, that's a little more difficult.

Basically, it all comes down to proprietary nanotechnology, if that helps. According to its maker and distributor, Ultra Tech International, Ultra-Ever Dry can be successfully used on almost any substance, from metal to plastic to cloth, and once the two-part system–a top coat and a bottom coat–has been applied, it works in temperatures ranging from -30°F to 300°F (-34°C to 149°C). If there's any limitation to the company's sales pitch, it's that the process was designed for work-related functions, which as of this writing has resulted in the company apologetically noting on its website that "we have not had the opportunity to test Ever Dry" for use on the bottom of boats, skis, surfboards, or snowboards. They might not have, but odds are that at least one of the 3.5 million people who've watched their video is surfing just above the water at this very moment.

CAR KITCHENETTE

Today, we do everything in our cars. Not because we like our cars so much that we never want to leave them, per se, but because the morning and evening commutes take so long that everybody does everything in their cars out of necessity. At least we can make the claustrophobic experience a little better with satellite radio, cupholders, and steering wheel-warmers. (And a little worse with all those empty coffee cups and fast food wrappers on the floor.)

But even in the very early days of the car industry, in the Model T days, there were a number of gadgets available to help a motorist spruce up their bare-bones Ford, including a dashboard flower vase, a detachable roof, even a kitchenette. Since fast food wasn't yet a thing (it wouldn't take off until American car culture did, after World War II) in the 1920s, a driver had to buy a Lincoln Kitchenette. You actually had to stop to use it–on roadside picnics and camping trips–but it was a refrigerated metal cabinet that stuck onto one of the car's running boards and folded out when needed to create a table. Inside, it held 25 pounds of ice, which kept food cold for up to 24 hours. There were also little compartments marked to hold "flour meal," "ice water," and "eggs," which probably didn't survive the trip.

AUTOMATIC HAT TIPPER

What ho, and good day, fellow gentleman. I see that your arms are quite laden with packages, overwhelming your spirited walk with various and sundry dry goods and whatnot. How then, I pray ask, are you to tip your hat to a passing lady, as custom requires?

Fortunately, our fellow goodfellow James C. Boyle has, in this very year, the olde-timey annus of 1896, registered with the United States Patent Office his "Saluting Device"! Bully! This most curious, although certainly useful, contraption attaches to the bottom of a hat (which you yourself must provide of your own volition via your finest haberdasher) and affixes to the top of the head. It is wound like a child's plaything, and all you need do is bow your head slightly. A pendulum sets forth machinery, of which this country is known to be the finest manufacturer, whereupon the hat spins, ever so slowly. But it does a full 360-degree turn and settles back in place. Oh, but what a sight it shall be as it spins about entirely upside down and back. The entertainment of the season! How we shall chuckle and guffaw and yet also be mannerly to the fairer sex!

THE BOYFRIEND BODY PILLOW

Hey, ladies.

Aren't human guys *sooooo* last century? They're always watching sports, they rarely pick up after themselves, and they fart so darn much. Isn't it time you ditched the irritating and disgusting male in your life? Now you can. Deluxe Comfort, a Detroit-based home product company, has released the Boyfriend Body Pillow.

The pillow, which is shaped like a man's chest and comes with a plush arm that you can wrap around your shoulders, will happily snuggle with you anytime you feel like it. Plus, the Boyfriend Body Pillow will never leave his tighty-whities on the bathroom floor, and it won't insist on watching a Golden State Warriors game rather than *Eat Pray Love*. Each Boyfriend Body Pillow comes with a removable microfiber shirt. The pillow is also machine-washable in case you accidentally spill some wine on it during a *Cougar Town* marathon.

Oh, and guys: You may now need to buy one of Deluxe Comfort's Girlfriend Body Pillows. It's basically the exact same thing as the Boyfriend Body Pillow, only it comes with a bosom made of memory foam.

ARAB AUTOMATA

An *automaton* (plural, *automatons*) is an automatic machine, often mimicking some human or animal action. There are many examples of such devices going back to ancient times, but the most impressive may be the inventions of the Arab engineer Al-Jazari, who lived from 1136 to 1206. Al-Jazari created about 100 automatons, including a 23-foot-tall water-powered clock that included the figures of a driver, a phoenix, and a serpent on the back of a life-size elephant. A series of contrivances sank a bowl into a bucket of water, which raised a seesaw, dropping a ball into the serpent's mouth, causing the serpent to tip forward and pull the bowl back up with a string. The same water-driven mechanisms triggered the arm of the elephant's driver to strike a cymbal every half hour, signaling the phoenix to chirp.

Al-Jazari also devised robot waitresses to serve drinks and hand-washing basins that were refilled by mechanical serving girls. But his greatest invention sounds like a tale out of *The Arabian Nights*: a proto-robotic band used to entertain guests at royal drinking parties. Four lifelike robot musicians were mounted in a boat and could be programmed to play different songs by changing the arrangement of tiny pegs in their inner workings. The musicians had over 50 facial expressions and bodily movements, which differed depending on which song they were playing. Not bad for a machine built more than 800 years ago.

THE HOLODECK

In 2012 the world came one step closer to science-fiction technology (or "Treknology," if you will) when Microsoft patented a virtual-reality installation reminiscent of *Star Trek*'s Holodeck—the magical rec room where Starfleet crew members interact with simulated people and objects in a perfectly realistic artificial environment, via sophisticated holograms.

Microsoft's version builds off of existing virtual-reality technology. Current VR systems require users to wear a special helmet or goggles to simulate a different visual environment. Microsoft's innovation will use multiple video projectors to shine images on all surfaces of a room, creating a 3-D illusion around the viewer in all directions. The space will appear to be completely transformed.

Even the furniture will seem to vanish; sensors will detect and compensate for any objects in the space, smoothing out visual distortions that would ruin the illusion. Motion detectors will track the viewer's movement around the room and adjust the projected images accordingly, maintaining a seamless, immersive visual experience. No word yet as to when Microsoft might bring this technology to market, but it's already gearing up for potential applications for video gaming.

Green alien go-go dancers not included.

ANTI-CARJACKING FLAMETHROWER

Carjacking is terrifying. You're just sitting in your car at a stoplight, when suddenly a thug runs up to your window with a gun and demands you get out and give him the car.

There aren't a lot of ways to prevent such a sudden, and ultimately speedy, crime. But in South Africa, where carjacking is epidemic, an inventor did figure out a way to punish carjackers before they get away with it: fire. Lots of fire. Charl Fourie introduced the Blaster to the South African car-peripherals market in 1998, at a price of 3,900 rand (about $655 back then). At the first sign of trouble, the driver activates the Blaster from the steering column. The Blaster squirts flammable liquid from a holding tank in the trunk out of two nozzles under the driver-side door. The Blaster then emits an electric spark, which ignites the gas…and the carjacker.

PROGRAMMABLE TATTOOS

The intimidating thing about tattoos—besides their outlaw associations—is that they're so darned permanent. Pity the fellow with his ex-girlfriend's name in ink. Barring expensive cover-ups or laser treatments, he's doomed to a future of awkward first dates, trying to explain away the "Jennifer" on his arm.

But what if tattoos were rewritable? What if changing that "Jennifer" to a "Susan" was as easy as sending a text message? Researchers are looking into different ways to create modifiable body art. There's already a patented process using magnetically-responsive inks to produce tattoos that can be redrawn with electrodes attached to the skin. Scientists at the Rensselaer Polytechnic Institute in New York take a different tack. Instead of using ink, they imagine a "tattoo" that's really a digital display—a thin sheet of polymer, encapsulating billions of carbon nanotubes, implanted under the dermis and shining out through the skin.

Uses for this technology might include displaying emoticons to indicate what sort of mood we're in, or jotting down messages where we know we won't lose them. (Shades of the movie *Memento*.) The screens could even display vital medical information.

Or they could help avoid the embarrassment of a souvenir that lasts forever—even when love doesn't.

GERBIL SHIRT

A gerbil's life is a lonely one, spent mostly in a cage, running on that wheel and waiting for those few minutes a day when its owner decides to play with it. Sometimes a change of scenery is needed for the animal's peace of mind, and for that reason, somebody invented the Gerbil Shirt.

With a series of tubes attached to both sides of a vest and two small chambers at the stomach, the Gerbil Shirt is basically a portable, wearable Habitrail. It allows gerbil owners to take their pet for a walk—just like dog owners do. The inventor even took safety into consideration. There are air vents throughout so that the animal can breathe easily, and the passageways have textured interiors so that it can maintain its balance, particularly on the vertical paths.

However, there is a serious downside. By allowing you to take your gerbil out into the wild, the Gerbil Shirt broadcasts to the world the fact that you are both a gerbil owner, and the kind of gerbil owner who wants and needs to take its gerbil out in public via a special shirt.

THE FLEXI-PHONE

At the 2013 Consumer Electronics Show in Las Vegas, Samsung shocked and delighted almost nobody by presenting prototypes of an extremely thin, bendable cell phone and tablet computer, part of a product line called "Youm." Instead of using sturdy materials like glass to build their Youms, Samsung opted to use a strong and flexible plastic. The proposed devices work by means of something called Organic Light-Emitting Diodes, which certainly sounds cool and futuristic.

Though indeed bendier than the average cell phone, the Youm phone isn't flexible enough to do anything potentially useful, like, say, fold completely in half so it can be placed into a pocket. It can, however, be rolled into a tube.

Brian Berkeley, VP of display research and development at Samsung (who, incidentally, quit his gig at Apple in order to move to South Korea for the job), demonstrated the purported benefits of a flexible tablet on stage. "We have expanded the canvas available for content," he said, pointlessly displaying a news-ticker-esque line of text scrolling across the edge of the tablet he had just bent in half.

THE FAKE TV BURGLAR DETERRENT

Uncomfortable keeping a gun around the house? Unable to figure out the buttons on one of those fancy alarm systems? The Hydreon Corporation has developed the perfect solution: FakeTV, a timer-operated device that tricks potential home invaders into thinking you're home when you're not. Crime prevention and TV: two great tastes that taste great together!

It's about the size of a coffee cup, but it emits an LED light pattern as bright as a big-screen television, with shifting colors and motion to simulate, when seen through a window outside the house, an actual TV in action. Sensors turn it on at dusk and operate it for four hours (if you feel like pretending to turn in early), seven hours, or all the way until dawn (if you want to make robbers think you stayed up all night for a *Growing Pains* marathon). No matter which setting you choose, you'll convince potential robbers–not to mention your neighbors–that 1) you're always home, and 2) you never stop watching TV.

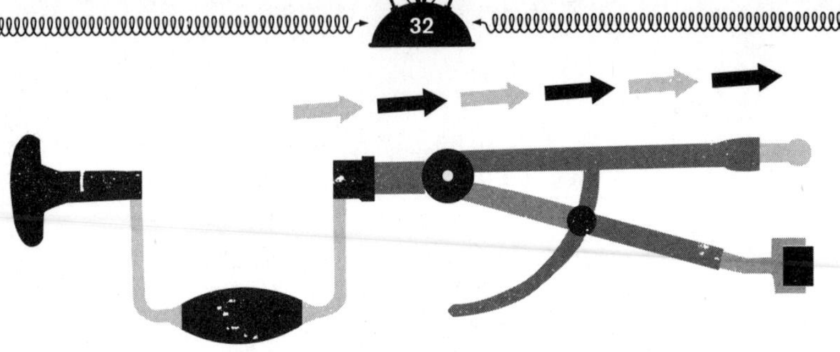

DIMPLE DRILL

Facial dimples—those indentations that accent a smile, making it at least 38 percent more adorable—are the result of a harmless deformity of a muscle in the cheek, the zygomaticus major. Usually, you have to inherit them from a parent. But what about those of us who lack such genetic good fortune? Must a dimpled cheek be born? Can it not be...made?

Inventor Martin Goetze thought so. In 1896 he patented the Dimple Drill. This contraption looked like a refugee from a woodworker's bench, sort of a gimlet mechanism crossed with a draftsman's compass. The dull tip, made of ivory or hard rubber, was applied to the cheek, and the handle cranked. The combination of gentle pressure and rotation would carve those darling divots into your mug in no time. The patent application claimed that the drill could be used to maintain and even deepen existing dimples, too.

And if boring craters into your face with a glorified auger seems a little extreme to you, well, nobody said that being adorable was easy.

BRAINPUT

As modern drivers, we all know the feeling of wanting to pay attention to the road but getting distracted by equally important activities like watching cat videos on our cell phones or eating a cheeseburger. The solution? Brain probes and robots, of course. Scientists at MIT are working on something called "Brainput" to help us multi-task more safely and effectively, because let's face it: We're never going to learn to actually pay attention to what we're doing.

The system works like this: You attach two probes to your forehead, and, using what is purportedly science but sounds an awful lot like diabolical voodoo, the probes sense when your attention is drifting and notify a robot to pitch in and help you out (e.g., your car drives itself for a minute so you can focus on licking a dollop of errant mustard off of the steering wheel). If the concept of robots reading your mind causes you to repeatedly scream and then run way, don't worry just yet: The technology has only been tested in a very basic form, in which the robot helps the user navigate a maze. Brainput's application in real-life situations, like the driving-whilst-texting-and-noshing mentioned above is still but a distant dream/nightmare.

AUTOMOBILE URINAL

Sometimes you gotta go when you're on the go, but you don't want to stop in order to go. Is there any way you can go while you go?

Yes, because Aston Waugh of East Orange, New Jersey, invented the Automobile Urinal. Anybody can pee into an old Snapple bottle; Waugh's device is fully functional, more than merely a pee holder you throw out later. It consists of three parts: a hanging water tank, a miniature padded toilet bowl that the driver sits on while driving, and a waste-storage tank that stows neatly beneath the driver's seat. After use, the driver flushes the device by opening a valve; water from the hanging tank flows through a tube into the toilet bowl, and from there into the storage tank underneath the seat. "For privacy," the inventor advises, "the user may wrap a large towel around him or herself from the waist to the knees before undoing the clothing to facilitate urination."

So you're free to go, so long as you planned beforehand, and you're not on a date or something. And you probably shouldn't use your cell phone while you're both driving and going to the bathroom.

A BETTER SKIPPING STONE

Despite the number of previously popular outdoor pastimes that've been deemed environmentally unfriendly in recent years, fishing and hunting among them, it still seems hard to believe that someone would have zeroed in on skipping stones as a seemingly fun thing that needed to be tweaked for political-correctness purposes.

But here's the deal: Technically, most rocks are inorganic. You throw them in a river or lake, and they just sit there forever, where they grow smooth over time via natural erosion. Inventor Glen Davis simply didn't think any of this would do, so he created the perfect stone for skipping, giving it "a smooth and continuous exterior surface with a substantially circular and smooth-edged outer perimeter" and "a hollowed, substantially elliptical and interior cavity," thereby resulting in a greater likelihood of some serious skipping action.

Presumably as a result of the sort of legality that causes wing-shaped pieces of chicken containing no actual wing meat to be called "wyngs," however, it must be noted that this handy-dandy item is technically not a biodegradable skipping stone but, rather, a "Water Skipping Article Incorporating Elliptical Outline and Hollowed Interior Core." Make it out of something biodegradable, and you're saving the Earth, one half-hearted stone's throw at a time.

4 WAYS TO STOP OVEREATING

Padlocking the fridge has been a favorite technique of the struggling dieter for decades, but you know what's even easier than looping a chain around the refrigerator door? Locking a cage over your face. That's the thinking behind the **Anti-Eating Mouth Cage**, anyway—a device that looks and operates pretty much exactly the way you might think. Looking like a cross between a catcher's mask and something Hannibal Lecter might wear to the beach, the Cage will prevent you from stuffing your face for exactly the amount of time you need to bust the teeny-tiny lock the designers were able to fit on its slender side strut…or however long it takes to drive to McDonald's, order a milkshake, and wedge the straw between the bars.

Don't feel like strapping on headgear but still interested in physically preventing your mouth from swallowing gross quantities of food? The **Dieter's Dam** may be just the gadget you're looking for. Unfortunately, it hasn't yet made the leap to mass production, but don't worry—to duplicate the experience, just take a few strips of semi-rigid plastic and tape them across your mouth, from cheek to cheek. The visual effect is kind of like a football helmet's facemask, only without the helmet or the pride in completing any sort of athletic activity. The emotional effect of having stretchy plastic taped to the face seems liable to leave the wearer too embarrassed to eat. Well played, Dieter's Dam!

If you're trying hard to lose weight but physically barricading your face isn't quite your style, don't fret: **The Alarm Fork** is here to help by inserting a nagging, motion-based sensor into your cutlery that goes off if you eat too quickly. It might sound kind of silly, but the idea is based on solid data; slowing your pace while eating gives your stomach a chance to tell your brain that it's full. And while the Alarm Fork's original 1995 patent didn't find any backers, the idea has resurfaced as the HAPIfork, a new $99 gadget that collects an assortment of information about your eating habits and sends it wirelessly to your computer, which then tells you that you eat too fast and need to lose weight.

Finally, for the smartfork owner seeking that last little bit of accessory oomph, we have the rather inelegantly named **Hand Near Mouth Alarm**, which combines all the rugged aesthetic appeal of Dick Tracy's watch with cutting-edge (for 1990) gyroscopic technology that senses when the wrist it's attached to is making mouthward motions too quickly. Of course, fooling the Alarm is as simple as not using that hand to eat every few bites. But on the other hand (get it?), constantly switching grips probably counts as some sort of exercise, so you're halfway there, diligent dieter!

SILENT-MOVIE SPEECH BALLOONS

Modern-day motion pictures are all about realism—actors speak words, and you hear those words at the same time they come out of the actors' mouths. The advent of this convention, which we take for granted today, was a big, reality-matching step-up from silent films, in which largely pantomimed, highly theatrical motions were interrupted by short frames of written dialogue, and only when absolutely necessary.

In 1917 inventor and frustrated silent-movie fan Charles Pidgin came up with a way to integrate dialogue into the moving image without "the use of a separate screen or picture or marginal sign." No, he didn't invent synchronized sound. He wanted silent films to employ the use of inflatable rubber tubes that had dialogue written on them. Pidgin literally invented the speech balloon.

Hollywood failed to use Pidgin's design; sound came along a decade later.

THOUGHT SCREEN HELMET

In 1998 former NASA technical writer Michael Menkin invented the Thought Screen Helmet, an aviator hat lined with Velostat, a kind of metallized plastic, and outfitted with secure straps. What does the Helmet do? It prevents aliens from reading your thoughts, and thus brainwashing you into allowing them to kidnap you.

Menkin doesn't actually sell the Helmets (you can make your own like he did; Velostat costs about $30 per yard). He just uses his website to tout the invention and explain why it's necessary. For example:

• "Aliens cannot immobilize people wearing thought screens, nor can they control their minds or communicate with them using their telepathy. When aliens can't communicate or control humans, they do not take them."

• "Adults and children all over America, all over Australia, in Canada, the United Kingdom, and in the Republic of South Africa are wearing Thought Screen Helmets to stop alien abductions."

• "Other shielding material was tried in previous models with less success. Only Thought Screen Helmets using Velostat are effective."

ELECTRONIC CIGARETTES

These ain't your grandparents' cancer-causing tobacco smoking sticks; these are "electronic cigarettes." Though a combustion-less non-tobacco cigarette was patented in the 1960s, it wasn't until 2003 that Hon Lok, a Chinese pharmacist mourning the loss of his father to lung cancer, came up with the idea for a safer alternative to smoking.

It's not really a cigarette; it's more akin, ironically, to the inhalers people with asthma use to breathe easier. The battery-powered device atomizes a liquid nicotine solution into an odorless aerosol mist that the user sucks out of the e-cigarette system. Resembling a traditional cigarette, its plastic body is complete with an LED light at the tip, that illuminates and dims with each puff, so as to look like a burning butt. Initially marketed as a smoking-cessation aid, e-cigarettes are merely a surrogate for nicotine inhalation and smoking simulation. Many question their safety. While they lack tar and other harmful chemicals found in traditional cigarettes, the FDA says they contain carcinogens and chemicals like diethylene glycol, which can be found in antifreeze. What a drag.

HYPER-REALISTIC FACE MASKS

From the country that brought us underpants in vending machines and Salty Watermelon Pepsi comes the perfect intersection of expensive narcissism and sphincter-puckering horror: three-dimensional, perfectly lifelike masks that look exactly like your face. Created by Japanese company REAL-f, the technology (dubbed 3DPF, for "three-dimensional photo forms") works by photographing a person's head from different angles, then printing the composite image over a resin that's been stretched over a mold. The finished product comes in two varieties: a face mask ($3,920) and a complete, mannequin-style replica of your head ($5,875).

Fake face masks are nothing new, but REAL-f justifies its extreme price point by taking things to a whole new level of detail–down to the irises and blood vessels–and the results are undeniably impressive (by which we mean "supremely creepy"). No one has ever created face replicas this close to the real thing.

Of course, that might have something to do with the fact that, as a consumer product, the 3DPF doesn't really have any practical applications beyond using it to terrify unsuspecting friends and family. That means it probably won't find much of an audience outside of people who have a dark sense of humor and too much money.

CASKET-VISION

When you think about caskets, what's your #1 concern, apart from the crippling, claustrophobia-inducing panic that comes from knowing you'll one day be placed into one forever? The fact that they're closed up forever and you don't get to see inside of it, or see your departed loved one, as they slowly decompose into a creepy skeleton, of course. Who wouldn't want to watch that?

With that ghoulish idea mind, in 1922 Jacob Fishman invented the Attachment for Caskets. It was, more or less, a reverse periscope, designed for looking inside of a buried casket and checking in on the decaying remains of a family member or friend.

At least the scope didn't stick out of the ground; it was telescoping, so it could be pulled out when necessary and put away when not. It was even outfitted with a light, because it's dark underground.

Fishman's design also included a lock for the eyepiece, so as to prevent rain, snow, and dirt from falling into the tube, or even onto the corpse inside. It also would prevent unauthorized looky-loos, because anybody who isn't a family member or a friend looking at your departed loved one is weird.

BUTTER STICK TYPE

Spreading butter on a piece of toast can be such a pain. You've gotta take the butter out of the fridge, get a knife, and struggle to get the butter on the toast even though it's cold and won't spread very well. But someone came up with a brilliant solution for this common frustration: the Butter Stick Type. The BST looks like a large glue stick but dispenses delicious, soft, easy-to-spread butter pats.

You would think that since this butter applicator actually debuted in the '90s, surely you'd be able to find one in the dairy aisle of every supermarket in America. There's no way that something so useful wouldn't instantly become popular, right? But the BST never caught on. In fact, it may have never been a real product at all. A photo of a BST, dating back to 1995, has been floating around the Internet for the past several years. It's popped up on countless blogs and Tumblr accounts, but no one seems to know where to buy one or if the sticks ever went on sale.

The most likely explanation is that the image was created for a *Chindogu* contest. Chindogu is the Japanese art of creating ingenious products that seem like an ideal solution to a common problem but are actually completely absurd. An editor named Kenji Kawakami came up with the word and later popularized it as part of a regular feature devoted to goofy inventions in his magazine, *Tokyo Journal.*

FART-ABSORBING UNDERWEAR

Australian company 4Skins is in the business of air purification, or maybe even environmental protection. It makes a product also called 4Skins, which utilize specially engineered space-age textiles to absorb, and thus prevent the public release and notice of, your stinky farts. The company motto, and product slogan: "Keep It in Your Pants."

4Skins' Contrast and Modern Classic lines of underwear are made using a technologically advanced fabric that incorporates something called "odor-eliminating nanotechnology" into every fiber. This futuristic fabric "attracts, isolates, and neutralizes" your bean-stinks immediately.

One question you're probably asking: Where do all those absorbed fumes go? Well, they go into the fabric, where they stay until you wash your 4Skins. The odor is then released into your washing machine.

Be forewarned that 4Skins absorb fart smells only, and not fart sounds.

BANANA SUITCASE

Your mother was right: Bruised bananas are perfectly fine to eat. But who would want to? The beautiful yellow skin of a banana marred by a big, ugly brown spot is quite unappetizing, even if it has absolutely no effect on the edible fruit inside the inedible peel. Except for, maybe, an extra-squishy spot. But it's a banana, so who cares, right?

We all care. Bruised bananas are gross. End of story. That's the thinking behind the Banana Suitcase. Bananas will inevitably take a beating in transit in a lunch box, brief-case, or purse, but this gadget keeps them pristine. It's a tiny, curved, banana-shaped yellow plastic box that's hinged in the middle.

Indeed, it does work to protect a banana...so long as the banana you've got is the exact size and shape as the Banana Suitcase. Otherwise, if it's too big, it's going to get bruised when you try to fit it in there. And if it's too small, it's going to bang around inside the case and get bruised. And if it's not the right shape, well, you're out of luck.

BEER CAN ROBOT

A can of beer rests on a table. A would-be drinker reaches out, anticipating frosty refreshment. But the top of the can extends upward, like a prairie dog peering out from its hole, then retracts, while three scuttling legs unfold from its base. Then the can shuffles away, leaving its victim frustrated and baffled, with his or her whistle tragically unwetted. It's like a *Transformers* cartoon gone horribly awry.

A booze-induced hallucination? Nope. This is CanBot, the creation of amateur roboticist Ron Tajima and revealed to an unsuspecting world in 2011. CanBot, controlled with a modified Nintendo Wii remote, moves independently, either by walking upright or by rolling. The body of the can is filled with batteries, control circuits, and the servo motors that give it motion. Meaning that even if you can catch it, there's no beer reward inside.

So why would anybody create such a thing? Tajima makes no claims to expanding the frontiers of robotics; CanBot is little more than an elaborate practical joke, perfect for scaring the bejeebers out of friends who've had a few beers already. In other words, it's a high-tech variation on the old "spring-snakes in the can of mixed nuts" gag.

BIONIC EXERSUIT

We can exhaust him…we have the technology…

In 1998, perhaps while buried under a pile of Thighmasters and Ab Belts and Bun Lifters purchased by phone during late-night infomercials, a young inventor decided the time was right for a single training outfit that would offer resistance everywhere, from head to toe. The idea was to turn every human movement into exercise, and every room into a gym, because the wearer would bring his workout with him, wherever he went.

Soon the inventor emerged from his sewing room with the Bionic Exersuit, a one-piece experiment in not-so-haute couture that wove taut rubber tubing from the neck down the arms and torso, and all the way to its attached footies. The Exersuit was blue and green and yellow, and ugly all over. Perhaps it was the outfit's less-than-glamorous looks that kept more workout mavens from experiencing the joy (and, no doubt, increased fitness) that could only come from that little extra tug of resistance in the fingers every time the user grabbed a pen…or in the knees when he sat…or in the chest when he breathed in and out…

PHONE GAS

Kids today are wild about drugs they aren't supposed to take, like OxyContin and Four Loko, but when you try to get them to take a puff on their life-saving asthma inhalers, they simply can't be bothered. Also, you can never get them to stop talking on their cradle handset corded phones. Right, parents?

Obviously that last part isn't true, but should we ever find ourselves plagued by these kids today who love landlines and hate asthma medicine, some enterprising soul has already patented a solution: Phone Gas. It sounds like technology that somehow hides flatulence in a handset until a prankster can let it rip on an unsuspecting victim, which would be awesome. Instead, however, Phone Gas accomplishes the rather mundane task of blasting the user in the face with aerosol inhaler medication via a specially designed phone-shaped object that is not actually a phone.

Instead of allowing the user to conduct a conversation with a human being who might be able to explain the cloud of mist wafting out of the microphone end of the handset, Phone Gas offers a "pre-recorded message or music playing into the earpiece" until the user opens his mouth ("in awe or boredom"), at which point a "magic button" is pressed, thus unleashing the life saving spray that the user has hitherto refused to inhale. Patented in 1988, it hasn't been a big seller, outside of *The Prisoner* re-creation societies.

THE LAVAKAN

Giving pets a bath is the worst—dogs hate water and are prone to jumping out of the tub when they're wet and covered in a mixture of grime and soap. Cats aren't any better; used to bathing themselves with their tongues, they just don't get it when you make them submit to a pool of standing water, and then they claw off all of your skin.

The Lavakan is an alternative to the heart-break of pet baths: It's a washing machine for cats and dogs. This industrial-strength machine soaps, rinses, and dries your pet in less than 30 minutes. The machine resembles a restaurant-grade dishwasher, or a tiny drive-through car wash, with a window so you can spy on your spooked pets. One of the inventors, Andres Díaz, claims that the 5-by-5-foot appliance can even reduce pet stress. "One of the dogs actually fell asleep during the wash," he said. Cost: $20,000, but if you have that kind of cash to spend on a pet washing machine, you probably aren't the kind of person who washes their own pets anyway.

THE FAKE FARMER

Much as the theory of relativity has remained a constant in physics, the universe of poultry farming has always had its own go-to formula: chicken + food = a plump and tasty chicken. But what if the poultry population isn't fattening up at a speed that's fast enough for a farmer's liking? Growth hormones may be the preferred method nowadays, but in 1981, someone came up with the idea of trickin' chicken into eating at a hastier pace with the Dummy Chicken Farmer.

Using the less-than-impressive IQ of the chickens to their advantage, this faux farmer may look like a human, but in truth, it's little more than a human-shaped bag of stuffing that hangs from a hook attached to its hat, traveling in a circular path around the chicken coop every three-and-a-half hours. It's like a scarecrow, but one that moves without scaring the birds. In an effort to both fake out and freak out the fowl, the Dummy Chicken Farmer also has an audio system in its chest that blares out whatever sounds the (real) chicken farmer has deemed most likely to inspire the birds to keep filling their faces. As if it weren't odd enough already, a bonus bit of inexplicability is the series of streamers which, based on the patent diagram, seem to be emerging from the dummy's nether regions.

BODY HAIR THINNER

When your formerly tight-cropped 'do starts turning into a full-fledged mane, it's clearly time to trim things back a bit. But when discussing the hair elsewhere on one's body, there's a bit more individual discretion as far as whether you're going to take just a little off the top or go the Full Monty and make with a wholly-hairless look.

That's where the Mudage Jolie Body Hair Thinner comes into play. Although described by the website Japan Trend Shop as "a handy device for men who don't want to scare girls away with their chest rug, but are reluctant to dispense with their body hair altogether," the Thinner is ultimately little more than a standard razor, featuring a stainless steel blade which helps with follicular manageability when run over your chest, underarms, arms, or legs. It's worth noting, however, that nowhere is it suggested that the Thinner is optimal for undercarriage use. That's not to say it couldn't be, but you'll, uh, probably want to take a bit more care if you go that route.

BEACH BOOTS

You've got to hand it to the mind behind Beach Boots, as it is a mind with the courage of its convictions.

See, the reason most folks go barefoot at the beach, despite the discomfort of walking across hot sand, isn't just because wearing shoes can leave your feet sweaty and gritty–it's because wearing shoes at the beach looks a little bit nerdy, like that time they photographed Nixon walking on the beach in a full suit, trying to look casual.

The inventor of Beach Boots, though, doesn't let that stop him. You can almost hear his creative process playing out: "So you think it looks nerdy, wearing shoes on the beach? Well, how about clunky plastic buckle-up clodhoppers tricked out with battery-driven motors and caterpillar treads?"

And with a click of the toe-controlled switch, you're tooling along the seaside in this unholy three-way hybrid of ski boots, roller skates, and a Sherman tank. Now, there are obvious utilitarian concerns: Will those mini-motors really propel the body weight of an average human? Is it wise to put spinning rotors so close to your extremities? And what's the matter with flip-flops, anyway?

Wear with knee-high black socks for full effect.

INHALED CAFFEINE

Making coffee every morning is a total pain. Even one of those newfangled individual-cup machines takes a few minutes to do its thing, and cracking open a can of Red Bull requires vital seconds when you're rushing to get out the door.

For those of us who can't stop hitting the snooze button and don't have time for a proper breakfast, let alone a latte, AeroShot Pure Energy could be the perfect morning eye-opener. It's the creation of David Edwards, a professor at Harvard University who has also designed other inhalable products. His first foray into this weird industry was Le Whif, a no-calorie, light-as-air brand of chocolate that goes straight to your lungs instead of your hips.

Each AeroShot inhaler contains eight puffs and 100 mg of caffeine, roughly the same amount contained in a tall Starbucks mocha. Another perk? These things won't give you stinky coffee breath. They come in three flavors: lime, raspberry, and green apple. AeroShot also has vitamins like niacin and B12, making each tube at least as healthy as a bowl of Frosted Flakes. The inhalers debuted in early 2012. Cost: $2.50 each (which is less than a tall Starbucks mocha).

EDIBLE FECES MEAT

Recycling really is a wonderful thing. Thanks to constantly evolving technology, we can make a better planet by converting old newspapers into product packaging, aluminum cans into automobile parts, and human feces into steak.

Yes.

It all started because of Japan's sewage problem. See, 127 million people living in a country roughly the size of Montana makes for a lot of dookie. So, the Tokyo Sewage service approached lab researcher Mitsuyuki Ikeda to devise a solution. And he came up with…poop steak.

The process to create poop steak involves the extraction of bacteria-spawned protein from human excrement and cooking it to kill off the harmful bacteria. The extracted protein is then combined with a chemical-reaction enhancer and put through an extruder to create the new meat product. This so-called meat is enhanced with soy protein (for flavor, naturally) and colored with red dye. The resulting "steak" is 63 percent protein and reportedly contains fewer calories than conventional meat, meaning the kind that didn't come from poop.

Currently the main obstacle to producing this new meat en masse is the high production cost. And the fact that most people have an aversion to eating poop.

SCARY CHILDBIRTH APPARATUS

Evidently childbirth wasn't scary and potentially dangerous enough, and it lacked an element in which a woman was strapped down to a table and spun around at high speed. So said inventors George and Charlotte Blonsky, who devised an "apparatus for facilitating the birth of a child by centrifugal force."

In their patent application, the Blonskys awkwardly and racistly say that "primitive peoples" have muscle systems so developed that childbirth is a breeze; not so for "civilised women, who do not have the opportunity to develop the muscles" within the realm of polite society.

Their apparatus aids in the delivery of a baby via 125 separate components, among them bolts, brakes, a variable-speed motor, stretchers, ballasts, clamps, a girdle, and more. In short, it uses repetitive, mechanical power and centrifugal force to both help the woman push out the baby and force the baby out of the woman in as timely and grueling an ordeal as possible.

Obviously, this invention was patented during the mechanics-obsessed and kind-of sadistic Industrial Age. But it wasn't. No–it was patented in 1963.

ROCKING-KNIT CHAIRS

For centuries, the greatest minds in engineering have struggled with the dilemma of how to effectively combine the boredom and discomfort of sitting in an oversize, wooden rocking chair with the tedium of knitting large quantities of wool hats. But now, after it seemed all hope was lost, we have the answer: the Rocking-Knit Chair.

Credit for the Rocking-Knit Chair goes to students Damien Ludi and Colin Peillex, who created it for the University of Art and Design's Low-Tech Factory Exhibition in Langenthal, Switzerland. The back-and-forth motion of the chair powers a set of gears that automatically draw yarn from a spool and knit it into a tube pattern. The knitted yarn can be used to make hats, large socks, and more hats.

Supposedly, the chair was meant to show how a relatively simple machine could perform relatively complex tasks, but we all know better. The Rocking-Knit Chair is obviously the first phase of an insidious master plan to enslave grandparents around the world and put them to work producing millions of useless items while making them think they are just relaxing.

INFLATABLE BREAST IMPLANTS

If you really want breast implants but just can't bring yourself to pick a size, science finally has an answer for your unique and indecisive vanity: the Hinging Breast Implant, a gadget that unites humanity's timeless breast obsession with the technology behind Reebok Pumps.

It works, so to speak, by inserting what looks like a teardrop-shaped accordion file into the breast with–and here's the crucial part–a nozzle poking out of the skin. If you're a man, you probably read that part in slack-jawed astonishment; if you're a woman, you're most likely wincing the way guys do whenever some poor dad gets whacked in the jimmy during an episode of *America's Funniest Home Videos*. Either way, you get the idea: The nozzle allows the Hinging Breast Implant owner to inflate or deflate according to her mood, outfit, or whatever reason a lady might possibly have for an impromptu cup-size adjustment.

Although a patent was issued for this way back in 2005, it hasn't caught on yet–perhaps because enough of us still remember Reebok Pumps–but in a world where bagel-shaped saline face implants are a thing in Japan (see page 200), anything is possible.

VIDEO GAME ACCESSORIES

Atari Mindlink. This bizarre controller for the Atari 2600 was announced in 1983, but was never officially released. Ads promised gamers the ability to play games not with a joystick or a wireless remote, but with *their minds*. Well, sort of–the Mindlink was a headband that connected to the Atari via infrared sensors that were supposed to pick up on subtle movements the player made with muscles in their head. (The accessory's designers came up with a game called *Mind Maze* that really was supposed to have been playable via ESP.) The Mindlink never hit stores. A presentation at the 1983 Consumer Electronics Show proved it didn't really work at controlling on-screen movements, and testers complained that the device gave them headaches.

Wii Car Adaptor. Nintendo's Wii system is a home gaming console that attaches to a TV. One of the Wii's many optional accessories is the Car Adaptor, a monitor that snaps onto the system for gaming-on-the-go in the (hopefully) backseat of a car. However, the Wii is played via motion capture, requiring players to wave their arms around, an activity that's not exactly conducive to road safety.

Wii Bowling Ball. The Wii is packaged with a disc full of sports games, including bowling. When the Wii was first released in 2006, YouTube quickly collected hundreds of videos of people accidentally throwing their wireless Wii remotes through their TVs, windows, etc., which led Nintendo to make wrist straps for the remotes. Nevertheless, CTA Digital manufactured this accessory specifically for Wii bowling—a plastic bowling-ball-shaped Wii remote. The instruction book helpfully advises, "Never, ever release the ball!!"

THE BOXING BUBBLE HEAD

The major occupational hazard of boxing is probably all of the punches to the head and face, particularly if you're a bad boxer who takes a lot of blows, as they can lead to brain damage. This invention, filed with the U.S. patent office in 1987, sought to make boxing safer, if not an entirely different sport altogether. The Bubble Head is a durable plastic bag filled with a clear, thick, shock-absorbing gel. The Bubble Head leaves the face completely unguarded and vulnerable, but protects the rest of the head, on all sides, with a barrier of goo-filled safety plastic.

Another feature of the Bubble Head: a touch-sensitive reservoir filled with red dye. This red dye would create a surefire way to score boxing by counting blows—every time an opponent punches the wearer in the head, the dye would be released into the clear fluid. At the end of the fight, the loser is the fighter with the most red dye in their Bubble Head bag (instead of another red substance on the canvas).

IMPLANTABLE CELL PHONE

There are two facts of life that seem to be increasingly irrevocable: that cell phones keep getting smaller and smaller, and that we, the cell phone users, are obsessively attached to our cell phones, what with their abilities to text, browse the Internet, take terrible pictures of ourselves, and shoot pigs at birds. (Science has even coined the term *nomophobia* to describe the fear of being without a cell phone—*nomo* being a contraction of "no mobile.")

Two students from the Royal College of Art in London have chosen to accept this reality, rather than try to pry the masses away from their handheld devices. Instead, they've made possible the next step in the cell phone size and addiction progression and invented the Audio Tooth Implant. In short, it's the bare-bones, working parts of a cell phone…surgically implanted into a cavity in a tooth.. It's a tiny contraption, about the size of a tooth filling, made up of a vibration device and a low-frequency receiver. Via your jawbone, it transmits sounds directly to your ear (the jawbone is an excellent transmitter of sound). All told, it allows you to receive calls, and then talk to people, inside of your own head, in total privacy. Bystanders will simply think you ate your Bluetooth.

ICE-CUBE PIGEON REPELLER

There are lots of ways to get rid of pigeons or other unwanted birds that hang around on your roof, doing who knows what degenerate bird activities. You can stick metal spikes up there, set up a speaker to play ultrasonic frequencies that birds find annoying, or cover the whole mess in slippery or sticky substances.

Or you can set up a mechanical arm to throw ice cubes at them. In 2011 Preston Jones of California patented his Bird Repeller. In form and function, it's a tiny catapult. Simply load the Repeller with water, plug it in, and let it make ice. The ice automatically flows out into a bowl and is then launched, via a spring-loaded arm, one story into the air, onto the roof, which apparently scares the heck out of birds. As a bonus, Jones note, ice is "environmentally friendly and will not jam gutters." As another bonus, it can be set up to shoot ice via a timer or remote control, just in case those lousy pigeons get wise and connect your presence in the yard with scary ice bombs.

STUFF THAT CLEANS ITSELF

Japan's Nippon Sheet Glass Company makes all kinds of windows—virtually unbreakable windows, windows that can withstand bullets. Big deal. They also manufacture a window that can clean itself.

The Cleartect line of glass products is coated with titanium dioxide. That's a photocatalytic material, meaning that it reacts chemically to light. When sunlight hits the glass, that unleashes a chemical reaction breaks down organic material on the window into smaller and smaller particles. The coating is also hydrophilic, so instead of forming droplets on the glass, rainwater forms an even sheet that flows down the window, taking the super-miniaturized molecules of dirt away with it. (Note: If it doesn't rain often enough, you have to hose down the window yourself.)

Moreover, scientists at Hong Kong's Polytechnic University discovered that titanium dioxide—the same stuff that's used for self-cleaning windows—can be used for clothes, too. When applied to cotton (no other fabric will work) the titanium dioxide breaks down dirt and other pollutants into smaller and smaller particles, the same way it does on glass. Sunlight and movement, they hope, will eliminate the dirt.

DIGITAL EAR CLEANER

One of the pros of cleaning your ears, if there are any, is that you can't see inside of your gross ears while you're cleaning them. The con is that you can't see inside of your gross ears while you're cleaning them, so you don't really know if you're doing a good or a bad job, or missing any spots, etc.

Mimikakis—reusable ear-cleaning devices—are popular in Japan, the land that gives us the King's Idea, a super-futuristic mimikaki. In your hand you hold the main element, an L-shaped piece that looks like a periscope, and like a periscope has a small viewing screen in one end (but digital!). At the other end is a lighted camera/stick made out of a composite of glass, stainless steel, and "anti-bacterial" resin. Shove that end in your ear canal and clean away, all the while watching your progress on the screen.

It's significantly more costly than a Q-tip at $90, but it's way cheaper than buying a surgical "snake"-style camera, taping it to a Q-tip, and jamming it in your ear.

THE ELECTRIC DOORMAT

When you think of exhausting physical tasks that clearly need some sort of mechanical assistance to make them more feasible for the average person, obviously the first thing on that list is going to be "casually wiping your feet before you go inside a building." Well, it was for Henry J. Ostrow of Palatine, Illinois, who in 1957 applied for a patent for an electric door mat, thus electrifying something that never, ever needed to be electrified.

Ostrow's invention featured "a plurality of brushes which are automatically actuated when a person steps on the device to remove dirt and other foreign matter from the shoes," and used pressurized air to dislodge any remaining undesirable material. The drawings that accompany the patent make the device look not unlike a conveyor belt, which stands to reason, given that the word "conveyed" is specifically used when describing the transportation of all the bottom-of-your-shoe nastiness into a nearby waste-storage system. Sadly, although it seemed as though it had the potential to literally sweep customers off their feet, the electric door mat never took off commercially.

PERSONALIZED ACTION FIGURE MACHINE

PARTY is a Japanese company that in 2012 unveiled a new twist on the photo booth: tiny action figures built to look exactly like you, from head to toe. Here's how "Omote 3d Shashin Kan" works:

The client sits or stands completely still for 15 minutes while multiple high-resolution cameras take pictures of every part of the body. (PARTY recommends wearing simply textured, solid-color clothing.) The images are then scanned into a computer. A month later, PARTY sends a figure to you, in your choice of 4, 7, or 8 inch height.

The company set up a public, reservation-only exhibition to produce dolls in Tokyo in 2012, and despite a ticket price of around $500, all viewing slots were filled up within days.

SOLAR-POWERED BIKINI

You would think that by now beaches would offer phone-charging and power stations much like the ones typically found in airports. Until the National Park Service and/or David Hasselhoff rectify this terrible oversight, there's always the solar-powered bikini.

This high-tech swimwear was created by Brooklyn-based designer Andrew Schneider in 2011. Each hand-stitched bikini is made out of thin, flexible photovoltaic film strips woven together with conductive material. USB ports are attached to the top and bottom pieces of the suit and can charge everything from iPods to iPhones while the wearer soaks up some rays. So technically it's not a solar-*powered* bikini, it's a solar-*powering* bikini.

The bikini is also safe to wear in the ocean (provided you remove any and all electronics connected to it before you take a dip). The suit's solar panels are moderated by a five-volt regulator that prevents any unfortunate shocks.

Schneider has also revealed plans for a pair of men's swim trunks, called the iDrink, that can chill beer.

THE DISAPPEARING DRESS

It's every teen boy's dream and every father's worst nightmare. If clothes are a way for a person to express themselves, then Studio Roosegaarde, a design agency in the Netherlands, has created the most expressive dress possible, as it relays information about the wearer's immediate, most intimate feelings. Dubbed INTIMACY 2.0, the dress actually turns transparent when the wearer gets, well, "excited."

The black (or white), otherwise conservative-seeming dress is made out of leather, opaque "smart e-foils," LED lights, and a few additional electronics. Those doodads can read the heartbeat of the person wearing it, and when it suddenly increases, the garment becomes "more or less transparent," according to designer Daan Roosegaarde. The faster their heart races, the quicker it disappears, thus sending a clear message of interest to any nearby visual stimulus.

The designer is also hard at work on what he's calling INTIMACY 3.0, a line of clothing for both men and women. This one will include a suit that turns clear when a guy wearing it begins lying.

HANDS-FREE SANDWICH HOLDER

It's probably an apocryphal story, but history says that the sandwich was invented when degenerate gambler the Earl of Sandwich, not wanting to leave the gaming tables to sit down for a proper meal and eat with a plate, a fork, and dignity, requested that a servant bring him a piece of meat between two slices of bread—thus the sandwich, named in his honor, was born.

The sandwich began its existence as a slapdash convenience. But even its compactibility and portability still isn't enough for some people—the people who would buy and use the Hands-Free Sandwich Holder. It resembles—and probably is just a slight reworking of—one of those hands-free harmonica holders that Neil Young wears so he can play harmonica while he plays guitar. You shove a sandwich into the designated sandwich-holding area (where the harmonica would go), then put the device around your neck, situating the sandwich in front of your gaping maw. (Note: All of these steps have to be done with your hands). This allows you to continue doing important things (video games, Twitter) without having to eat with your hands like an adult human with self-respect.

EXTREMELY USEFUL KITCHEN GADGETS

Hutzler 571 Banana Slicer246: It looks like a banana with vertical blinds running through it. That's because this yellow, oblong (banana-shaped) gadget is made, specifically and only, for slicing bananas. Simply press it over a banana, and boom, banana slices. (You still have to peel the banana first.)

Mr. Marinator: This countertop gadget borrows its formal "Mr." address from the actually revolutionary Mr. Coffee. Instead of marinating meat in a bowl or on a plate, place a roast or large amount of meat in Mr. Marinator, along with your chosen sauce. Then Mr. Marinator shakes and agitates the meat with the sauce so it's marinated and ready to cook in under an hour.

Milk Carton Holder: Milk has come packaged in sturdy paper cartons for more than a century. It's very easy to hold a milk carton. Still, many companies offer a small plastic or metal "carton holder," basically a handle, for even easier pouring.

The Egg Cuber: It's a square-shaped cutter. You put it on a hard boiled egg. It cuts a square shape out of the egg. That's all.

DOG DNA TEST KIT

Not all of us are keen on taking our dog onto *The Maury Povich Show* to find out if that firecracker of a Schnauzer down the street is its biological father. Nor is it easy to get a dog to sign up with Ancestry.com. But when your dog isn't looking, you could swab the inside of its cheeks. Mars Veterinary sells Wisdom Panel dog-DNA test kits so owners can find out the make up of their canine's breed, in case it isn't abundantly clear.

Why the need to know? So doting pet owners can tailor diet, environment, and fitness programs to the dog's specific needs. (Coincidentally, Mars Veterinary also sells premium kibbles and pet-care products.) During development they analyzed more than 19 million genetic markers.

Will canine units on police squads soon be using these kits as forensics on other dogs? Not quite, but dog owners can continue to build on the imagined history of their pet, and when they do voices for them, they'll know which accents to employ.

EARTHQUAKE HOUSE

Anyone who has ever lived through an earthquake (or even seen footage of the aftermath) knows the destructive power that quakes unleash on buildings. But while geologists, engineers, architects, and other general egghead types try to build more quake-resistant houses, they seem to have missed out on one brilliant design: the Earthquake House.

Instead of the traditional four walls-and-foundation scheme, the Earthquake House is round and equipped with seismic sensors. When a large enough temblor hits, the house is automatically detached from its anchors, tethers, and utility lines, and is free to roll down the street or hillside with the tectonic punches.

The patent for the Earthquake House describes it as "Mother Nature's giant bowling ball," which sounds fun except for the fact that no one wants to live inside a bowling ball, nor did Mother Nature create the Earthquake House. Perhaps a more appropriate description might be "Mother Nature's giant hamster ball"?

But how does someone actually live inside a house designed to roll on the ground? The Earthquake House is designed with a self-righting inner living structure, so you don't have to worry about nailing your furniture to the floor. As for where to hang the Christmas lights, you're on your own.

CHEESE-FILTERED CIGARETTE

Okay, smoking is dangerous business no matter how you look at it. So for decades, cigarette companies, sensitive as they are to public health concerns, have experimented with different filter materials to make the smoking experience "safer." They tried charcoal, cork, and even asbestos (oops) before ultimately settling on cellulose acetate (a.k.a. plastic) around the mid-1950s.

Feeling that traditional filters did an inadequate job of removing tar from cigarette smoke, however, one inventor patented a novel approach in 1966—the cheese-filtered cigarette. The filter (more of a recipe, really) is made simply by grating a hard cheese (such as Parmesan, Cheddar, or Swiss) into small pieces and mixing it 2:1 with charcoal. The idea behind the mixture is for the cheese to more effectively filter out the nasty stuff, and for the charcoal both to absorb the cheese oil and to keep the cheese smelling and tasting its freshest without passing on any unctuous cheesy taste to the smoker.

Oddly, the inventor of the cheese-filtered cigarette was silent on what to do about the almost certain elevation in cholesterol smokers would face from using it. We recommend taking your anti-cholesterol medication with a creamy, ranch dressing-based dipping sauce.

THE BABY MOP

Babies drool, babies poop, babies scream bloody murder. Plus they're lazy, just lying there doing nothing all day. Fortunately, there's now a way to get your baby to help out around the house. In 2012 Better Than Pants, primarily a novelty T-shirt company, started selling the Baby Mop. It's more or less a traditional baby bodysuit with mop heads attached to the arm and leg holes. Plop a little one into the Baby Mop and they'll polish any floor as they crawl around. The item, which comes only in blue and has cartoon dinosaurs on the front, costs $40.

The outfit supposedly teaches babies a strong work ethic while they get exercise and burn up excess energy. Better Than Pants' inspiration for the product? A fake commercial for a Baby Mop from a Japanese sketch comedy show.

HELMET BAR

In the 1980s, a hat with cup holders and straws coming out of both sides was all the rage in stadiums. It gave sports fans a chance to applaud a good play without having to put down their watered-down, overpriced beer or, as the case was, *two* watered-down, overpriced beers.

The "beer hat" was such a hit that somebody took the idea one step further and created the Helmet Bar, a device for rabid sports fans who can't watch a game without consuming mass quantities of alcohol, but prefer classy mixed drinks to beer. The Helmet Bar, for which a patent was issued in 1987, held four bottles, which, when an assortment of valves across the front were properly opened, moved the liquid via tubes into a mixing chamber. The cocktail then flowed via another tube down to a mouthpiece, from which the world's laziest bartender could sip the now-mixed drink.

The contraption failed to make a dent in the public consciousness, presumably because either a) it looked ridiculous, b) it was too difficult to operate and fans decided to buy a Coke and sneak in airplane bottles of rum, c) the weight of four bottles on the head caused the neck to collapse, or d) the engineer who created it forgot to include a way to transfer mint leaves, maraschino cherries, and other crucial garnishes.

THE LEVITATING TABLE

Most dining tables just don't have enough leg-room. Or, more precisely, they don't have room for all the legs that end up underneath them: legs of diners, front legs of chairs, and of course the legs of the table itself. Remove those last four from the equation, though, and you've got more room to stretch out in comfort. Oh, yeah–and you've also got a freakin' levitating table, a dining surface just floating in space, in the middle of the room, with no visible means of support, other than *magic*.

It's not magic, it's magnets. This marvel, created by Belgian designer Yana Christiaens and inspired by European high-speed trains, uses principles of magnetic suspension. Powerful electromagnets mounted on the underside of the table keep it hovering over a steel plate below. Because the plate is set permanently into the floor, the table will be stationary, so pick your spot carefully.

On the plus side, the height of the table is adjustable via a control panel (recessed into the tabletop) that varies the strength of the magnetic repulsion current. Also on the plus side, this is basically the coolest thing ever.

THE PERIODIC TABLE TABLE

If you're like us (you're like us), then you're always rummaging around the house, looking for one of the 118 chemical substances on the periodic table of elements and wish that there were some easy to use system for keeping all of your elements, from hydrogen to lawrencium, safely stored in one place.

In 2002 scientist Theodore Gray of Illinois came up with the solution we've all been clamoring for: the Periodic Table Table. He built a three-dimensional Periodic Table of Elements Table–it's conference-table-sized–and included more than a hundred drawers. Each element group (alkali metals, noble gases, etc.) is represented by a different type of wood. He then filled the drawers with samples of as many real elements as he could get a hold of (sorry, no plutonium).

"One evening while reading *Uncle Tungsten* by Oliver Sacks, I became momentarily confused," Gray explains. "He begins a chapter with a description of a periodic table display he loved to visit in a museum, and in misreading the paragraph, I thought it was a table, not the wall display it actually is."

WHERE'S MY HOVERBOARD?

In a famous sequence in 1989's *Back to the Future Part II*, Marty McFly (Michael J. Fox), trapped in the futuristic year of 2015, flees a gang of hooligans by flying around on a Mattel-made hoverboard—a flying skateboard.

On an NBC special promoting the film, director Robert Zemeckis claimed that hoverboards are real. "They've been around for years," he said. "It's just that parent groups haven't let toy manufacturers make them." He was joking, but the switchboards at Mattel were reportedly overwhelmed by parents looking to have a hoverboard under the tree for Christmas 1989.

While the boards used in the film look convincing, they were actually rudimentary props. To film the sequence, Fox's sneakers were drilled into the hoverboard and he was suspended by cables to make it look like flight. Still, much like flying cars, the hoverboard become an iconic portent of the future, leading many to wonder "Where are the hoverboards?" Several attempts have actually been made to make them a reality.

• On a 2004 segment of the Discovery Channel series *MythBusters*, the show's crew built what they called the "Hyneman Hoverboard" (named for co-host Jamie Hyneman). It was constructed out of a surfboard and was propelled by a leaf blower's motor. It didn't work very well.

• In 2007 the crew of the British series *The Gadget Show* created another leaf blower-powered hoverboard...that couldn't be propelled or steered. Their second version added a small jet engine to the mix. It was more maneuverable but still barely functioned.

• That same year, a company called Future Horizons released a series of do-it-yourself hoverboard kits. The boards were made out of fiberglass shells, basically worked like miniature hovercrafts, and weighed 80 pounds. They were operable only on completely smooth surfaces and, in short, were a long way away from the movie version.

• A company called Arbortech released an actual, working hoverboard in 2009. Pros: It could float over both land and water, and reach a top speed of 15 mph. Cons: It was over six feet long and weighed more than 200 pounds.

• In 2008 French artist Nils Guadagnin began work on a more traditional hoverboard for an exhibition. His version was much more like the one in the movie, and it could float a few inches off the ground via electromagnets. A built-in laser system also helped keep it stabilized. But while this one looked cool, it couldn't support any actual weight.

• To date, the most successful prototype has been the one constructed by researchers at Paris Diderot, a French science and medical university. Unveiled in 2011 and called "The Mag Surf," this hoverboard can carry up to 220 pounds...and levitate just 3 cm off the ground.

DOG TRANSLATOR

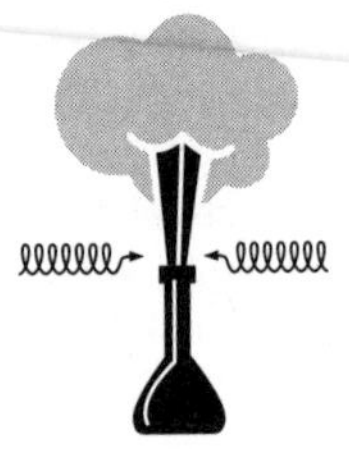

Researchers say that cats emit dozens of different sounds, and if you pay attention, you can ascribe clear meaning to each individual yowl, meow, and purr. A dog's vocabulary, however, consists almost entirely of "growl" and "bark." With the former you know he's upset; the latter could mean anything.

Translating the language of dogs, or wondering "what dogs would say if they could talk," has long been the subject of humorous science fiction, but better communication with a dog is now a reality: Introducing the Bowlingual Dog Translator.

Simply get your dog to "speak" into the gadget, which looks like and probably was at some point a walkie-talkie, and the Bowlingual will display one of six digital dog-face icons, representing "happy," "sad," "on guard," "showing off," "frustrated," or "needy." It's adjustable to the specific barks of more than 80 popular dog breeds, and is customizable according to "the size of the dog and the shape of its nose," because nose shape definitely affects the context of your dog's carefully chosen bark.

EARLY GPS

You probably rely way too heavily on your GPS to get you around town, and have done so, since, what, the late '90s at the very earliest? It turns out there was a predecessor to consumer-grade in-car GPS or "global positioning systems," and it dates back to the 1930s.

The product was called the Iter Avto. It did not talk to you in a calming female voice, but it did attach to the dashboard of a car, like a gigantic Garmin. Rather than feed satellite maps into your car, since they didn't exist at the time, the Iter Avto came with scrolled paper maps that you loaded into the machine. It was then hooked up to your speedometer, and the speed of your car controlled the speed at which the map advanced. It's sort of like a player piano that plays maps.

We don't know much about the Iter Avto, as the automatically-advancing map system, which appears from photos to be Italian-made, didn't really catch on. Perhaps because if you took a detour, the map would become almost useless. Or because a map had to be custom-made for a route. Or maybe because it didn't talk to you.

AN EARLY ELECTRIC RAZOR

This was one of the first handheld electric shaving razors, but the technology is a bit different than what we have today. In 1917 a battery-operated razor filled with hundreds of tiny but safely rotating blades would have made it prohibitively expensive, if not impossible to create, so Frank White did the best he could with the tools available at the time.

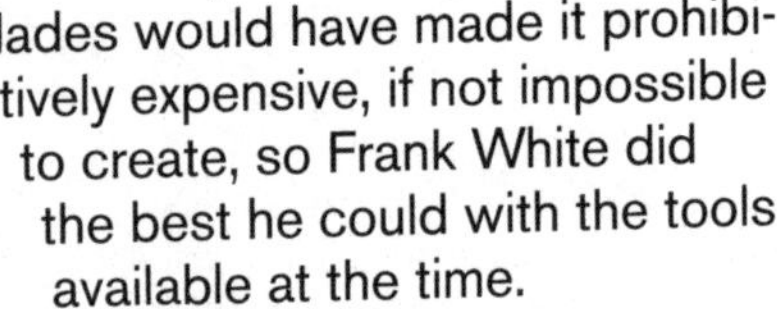

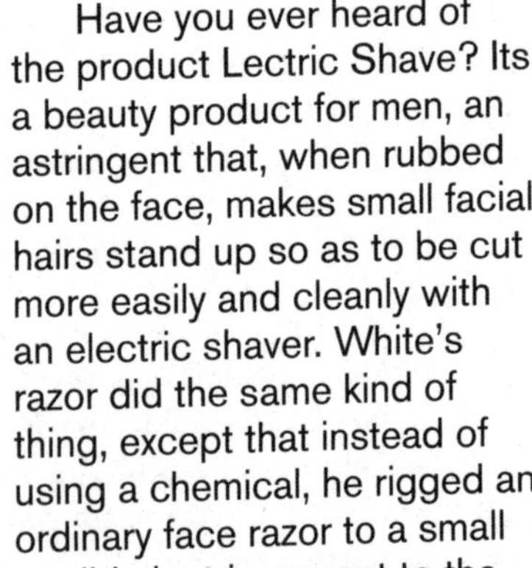

Have you ever heard of the product Lectric Shave? Its a beauty product for men, an astringent that, when rubbed on the face, makes small facial hairs stand up so as to be cut more easily and cleanly with an electric shaver. White's razor did the same kind of thing, except that instead of using a chemical, he rigged an ordinary face razor to a small battery. The battery sent a mild electric current to the shaver's face, which raised up the hairs so they could be lopped off neatly.

THE THUMB-SUCKING STOPPER

Proving that there are statistics available for everything these days, statistics show that businessmen rise through the ranks in a far more expedient fashion once they've managed to break the habit of sucking their thumbs.

Those with a vested interest in climbing the corporate ladder, even those who aren't in business but who are adults who still somehow suck their thumbs, may therefore wish to consider a thumb-sucking inhibitor which was patented in 1987. Given that it's formally described in its patent as an "Apparatus for Inhibiting Digit Sucking," the device also works on toes, if that's something you do.

Please note, however, that the device may prove somewhat unwieldy for those whose jobs require them to man a keyboard all day. As the patent explains, "A bracelet fits around the wrist, a primary ring attaches to the bracelet, (which) is supported by two or more tabs extending outwardly from the ring to the bracelet, and cross-tabs extend between the outward tabs to prevent unwanted withdrawal of the thumb from the ring." But it's all still probably easier than trying to type with a thumb in your mouth.

PORTABLE LADY URINAL

It's a problem as old as time itself—you want to pee standing up, but lacking the convenient, outer plumbing of a man, all you have between your legs is cumbersome, retreating lady business, necessitating a sit-down. The solution: GoGirl, a silicone funnel that a woman can hold against her crotch so that she can pee like a man.

The GoGirl website refers to the product as a "Female Urination Device" that "fits easily in your purse, pocket, or glove compartment." The company recommends it for all kinds of modern women on the go (as it were), be they cross-country road-trippers, backcountry skiers, or just germ-phobic mothers who don't want their daughters to sit down on public toilets. After all, nothing is more hygienic that peeing into a silicone tube and sticking it back into your purse, pocket, or glovebox.

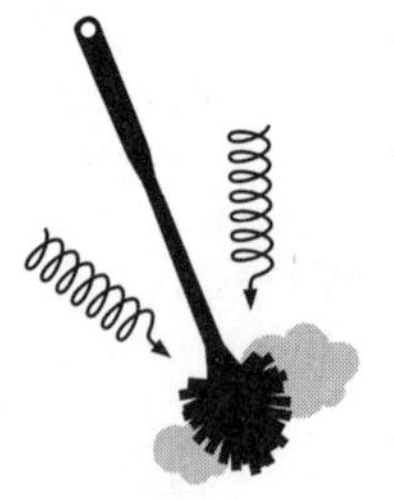

GoGirl is available in purple for girly-girls or khaki for more outdoorsy types.

THE WRIST GUN

Patented in 1929, the Automatic Concealed Firearm for Self-Defense was a gun that secured to the inside of the wrist with leather straps. It was concealed from view beneath the sleeve of a shirt or coat. A pull chain extended from the trigger to a ring worn on the ring finger, enabling the wearer to fire the gun with a backward snap of the wrist.

"Such a hidden firearm will be especially valuable in case of a holdup where the intended victim, when commanded to hold up his hands, or even before such a command, may shoot at the criminal without any further preparation, automatically when lifting his arms," inventor Elek B. Juhasz wrote in his patent application.

And if the robber has an accomplice? Juhasz designed a second version with two guns strapped to the arm, activated by a single chain. One pull on the chain fired the first gun; a second pull fired number two. Just don't ever catch the chain on anything.

HYPOALLERGENIC CATS

Scientists don't spend all their time doing stupid stuff like curing diseases and finding sources of clean, renewable energy; they've also figured out a way to create a superior house cat.

After their dismissive attitude, hungry wails at 5 a.m., and inclination to tear and scratch everything you own, the worst thing about cats is that lots of people are allergic to them. Well, technically, they're allergic to cat dander, tiny bits found all over the cat and where it lives, but you can't have one without the other. If you want a cat and are allergic to cats, you either take expensive medication or just deal with the constant sniffles and watery eyes.

Or you spend your life savings on a bioengineered, hypoallergenic cat of the future. Delaware firm Lifestyle Pets offers consumers the chance to buy the "Allerca," a cat whose genes have been tampered with so as not to produce the protein that triggers cat allergies in humans. Cost: between $6,000 and $29,000.

THE ELECTRIC PICNIC

First, get out there to the lake and catch yourself some fish. Second, don't forget your **Electro-fishing Pole**—an electrified stainless-steel loop with an insulated fiberglass handle. The user wears a battery-backpack, which is connected to the loop and has another wire in the water, completing the circuit. When a fish swims within the electric field created by the electrodes, ZAP! The fish succumbs to a minor seizire, leading it to quickly lose consciousness and become immoble, allowing it to be easily plucked from the water. We recommend using rubber gloves, both because it's a nearly dead fish and because of all that electricity mixing with water.

Next, clean and cook your fish in the normal, time-honored, non-electric ways. Finally, when you're ready to serve, bring out your **Electrified Tablecloth**. It's got a pair of built-in electrical strips powered by a 9-volt DC battery. Why does this exist? To repel insects and other picnic pests. An insect trying to cross the strips will get an electrical shock strong enough to discourage further travel across the table, in that the shock almost kills it, making the world safe for shocked-fish fricassee. Good news: The strips are not powerful enough to shock a person who accidentally touches them.

GOLF GADGETS

Golf-Swing Glasses. Mimicking the blinders worn by racehorses to keep them from being distracted by other horses, these glasses literally force the golfer to keep his eyes on the ball by staring through two "eye tunnels," because that's the only place they can look. If the golfer's head moves too far to one side, the ball disappears from view behind the thick, opaque ridges that surround the eyeholes.

Golfer's Crotch Hook. This device keeps a golfer's head down during a swing, where it should be…the hard way. The golfer wears a tight-fitting headband attached by an elastic cord to a massive, seven-inch fishhook fitted into the crotch of the golfer's pants. If the wearer lifts his head up too high during the swing, he is instantly–and uncomfortably–notified.

Putt Teacher. It's a putter attached to a belt on rollers (it looks sort of like a combination golf club and belt sander). If the swing is crooked, the belt will twist into a mess, telling the golfer what they probably already know, because they bought something called the Putt Teacher: that they're putting needs work.

Talking Golf Ball. A golf ball you can find in the rough! Inside the ball is 1/50 of a gram of radium; the golfer uses a handheld Geiger counter to locate the ball and retrieve it. But because this is plutonium we're dealing with, the manufacturer couldn't use enough to make the ball detectable from more than five feet away.

PILLS THAT MAKE YOUR POOP GOLD

There's a tradition among the super-rich of treating money with contempt. When a plutocrat lights his cigar with a $100 bill, he's brazenly demonstrating that he has, in the most literal way possible, money to burn.

Gold Pills take conspicuous consumption to a new level. These pills—standard gelatin pharmaceutical capsules, coated and filled with 24-karat gold—will allow high rollers to literally flush money down the toilet. When swallowed, Gold Pills, according to the makers, "turn your innermost parts into chambers of wealth." And as the indigestible gold leaf exits your system, it also gives you glittery poop. Fabulously expensive glittery poop, in fact—Gold Pills retail for $425 per dose.

Devised by artist Tobias Wong, Gold Pills were originally created for a 2005 gallery show called "Indulgences." The exhibit showcased items intended to satirize "our obsession with fame, celebrity, and commodities." Ironically, there was enough demand that Gold Pills were made available for purchase. Most are bought by art collectors, certainly. But there's bound to be some wealthy nitwit with more money than sense who's actually swallowed the things to show off his gold-encrusted leavings.

MICK FLEETWOOD'S DRUM SUIT

Depending on what sort of chemicals you've got in your system, drum solos are either the most boring part of rock concerts or the most awesome. Fleetwood Mac drummer Mick Fleetwood occasionally livens up his extremely long drum showcases with what he calls a "magical drum suit."

Fleetwood debuted the suit (really just a vest) during the band's 1987 "Tango in the Night" tour. Fleetwood would step out from behind his drum kit and make his way to the front of the stage. While another drummer performed in the background, he would hit various electrical pads sewn into the vest. Some played drumbeats while others emitted howls, car horns, or other sound effects. After several minutes of dancing around while slapping himself, Fleetwood would return to his kit and wrap up the solo with his conventional drums.

Fan reactions to the suit were mixed, but it helped distract from the fact that Fleetwood Mac's longtime front man and guitarist Lindsey Buckingham wasn't present for the tour.

DAN HARTMAN'S GUITAR SUIT

Glam rock was in vogue among musicians in the 1970s—Elton John and David Bowie and their costumes could have inspired at least a dozen sci-fi and fantasy films between the two of them. But one of the more inspired—and functional—outfits from that era of music history has got to be Dan Hartman's guitar suit.

Hartman was the bassist for the Edgar Winter Group, the blues-rock band that had hits with the songs "Frankenstein" and "Free Ride," which Hartman wrote and sang. He also had a hand in designing his $5,000 wearable guitar outfit. The one-piece jumpsuit featured a bass-guitar neck attached to the side with strings over the abdomen, along with cordless pickups and amplifiers sewn throughout. (Sounds were relayed via radio signal to the mixing board.) Hartman told reporters that he could feel the music, as every note reverberated through his stomach and the rest of his body as he plucked the strings.

It definitely blurred the line between "clothing" and "musical instrument" and looked like something straight out of *Starlight Express*. In addition to being difficult to get through doorways and into bathroom stalls, the guitar suit was probably not the safest outfit ever designed. Let's just say it's not the sort of thing you'd want to be playing during an outdoor concert when a thunderstorm breaks out.

THE IGROW HAIR HELMET

The $695 iGrow is a laser-powered helmet with hair-follicle-stimulating infared lights. As depicted on the models in the SkyMall catalog, it resembles an oversize bike helmet, with ear-covering headphones, that rests just above the head (to give the lasers room to do their magic), like an old-timey hair salon dryer. And those headphone-like things are actually headphones, so you can listen to music by plugging in your iPod while 51 laser/LED lights shoot beams at your receding hairline for an hour.

The validity of manufacturer Apira Science's claims that this gizmo regrows hair is unproven, although the company claims that the iGrow was previously "sold exclusively to doctors for use in hair clinics around the globe," meaning not the highly-regulated medical community of the United States.

SAFER RUSSIAN ROULETTE

As popularized by such movies as *The Deer Hunter*, Russian Roulette is a game that involves placing a single bullet in a gun's chamber, spinning it, then putting the gun up to your temple and pulling the trigger. You win if you don't blow your head off. We can all agree that Russian Roulette is a super-fun party game, but the only problem with it is getting the same group of good-time Charlies back together to play it again and again, because, as per the rules, somebody often dies mid-game.

Inventor Steve Zuloff came up with a way to make Russian Roulette fun for the whole family, and for people who don't want to shoot themselves in the brain. His patent for "Game Device and Method Thereof" is a way to make Russian Roulette not deadly. A balloon is squeezed into a ring that attaches to the barrel of a toy gun. Each player takes a turn by holding the balloon part of the device up to their head, and then they pull the trigger. The gun then shoots either nothing (winner!) or a pin into the balloon (loser!).

Everybody lives! Unless, of course, the gun malfunctions and you shoot a pin into your ear or something.

INVISIBILITY CLOAK

Since time immemorial, nerds have dreamed of using futuristic, fantastical inventions in real life: X-ray glasses, jet packs, and the invisibility cloak, which, among futuristic gadgets, has always seemed the most out of reach. And yet a Canadian company called Hyperstealth, which specializes in high-tech camouflage technology, has developed a material that confers genuine invisibility upon the wearer. The product, Quantum Stealth, works by bending light rays to hide whatever's underneath it from both the naked eye and infrared light. Just imagine all the geeky, pervy, and almost certainly illegal things you could do with a Quantum Stealth!

So can you get one before the first day of school at Hogwart's next fall? Sadly, Hyperstealth repeatedly states on its website that all photos and product demonstrations of Quantum Stealth are mere mock-ups, and that "for reasons of security" they can't discuss details of how the technology works or show images of it in action. So at best, the cloak is nowhere near being ready, and at worst, it's a complete hoax. It may not matter ultimately, as Hyperstealth plans on offering Quantum Stealth only to militaries, and not to consumers. Sorry, Gryffindors.

KEG HEAD

There are a lot of novelty gifts out there that encourage the beer enthusiast to drink more beer: the classic beer helmet (plastic baseball-style cap that holds two beer cans, with a straw running down to the wearer's mouth), and a thing we saw in a catalog, called the Beer Belly, a thick plastic bag that's worn under a shirt and is outfitted with a straw that runs through out the collar to the mouth.

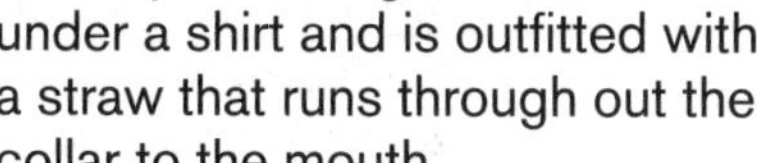

The Beer Belly (and its sister product, the Wine Rack, an alcohol-reservoir/brassiere) is designed to bring alcohol into places where alcohol isn't allowed. The Keg Head, patented in 1999, is the complete opposite of that. It's a hat, but it's also a functional mini-keg of beer. Resembling an old-fashioned wooden keg, it comes complete with a tap, allowing the user to reach up in front of their face and fill up a glass with a fresh draught. Have you ever lifted even a small keg of beer? Would you want one sitting on your head? It's not comfortable and would be quite heavy, but one side of the otherwise round Keg Head is tapered to fit firmly.

CAT EARS FOR HUMANS

One day, Japanese manufacturers will achieve their goal of turning every human being on the planet into the cutest robot/cartoon character ever. Until then, we'll have to make do with devices such as Neurowear's Necomimi Brainwave Cat Ears.

The device, which is basically a headband with a pair of fuzzy, pointed ears, has a forehead arm to detect neural activity (just go with it) and an earlobe clip to complete the circuit. In addition to making the wearer look like a telemarketer at a sci-fi convention, the Cat Ears are designed to read brain waves and convey the user's neural activity. (Because actually showing emotions with your face and stuff is hard work, and really personal.)

Clearly, the engineers at Neurowear considered the entire range of human emotions and then decided to just go with just four. So when the user is concentrating, the ears perk up. When the user is highly interested in something, the ears perk up and wiggle. When the user is relaxed, the ears droop. When the user is both highly focused and relaxed, the ears wiggle up and down.

And when the user gets mad that they dropped $70 on a piece of gimmicky garbage, the ears twitch in a trash can.

TEETH EXERCISER

You may not have noticed, but our society is in the midst of a dental crisis. Thanks to modern methods of food preparation, our cooked foods require little in the way of mastication (that's fancy-talk for "chewing"). The result? Rotted teeth, underdeveloped jaws, and just generally poor self-image.

Or at least that was what Charles G. Purdy believed, which led him to invent a device to reverse this pernicious trend. Purdy's patented teeth exerciser is basically just a mouth plate attached to a spring, which itself can be attached to a firm support (such as a wall) or to an exerciser plate in another person's mouth. The wearer then moves their head back and forth, and the resulting stress from the spring causes tension in the wearer's mouth. This stress is supposed to exercise the normally withered and useless muscles of the mouth and head, leading to improved oral health.

This sounds like a fairly sound concept, except for one glaring oversight. Based on what we've observed over the years, if there is one thing most people are already exceptionally good at, it's ensuring that their mouths get plenty of exercise.

SPRAY-ON WI-FI

Everyone is permanently affixed to their cell phones these days, and when you come to the rare place without reception or Internet data access, it's maddening. And it's a growing industry—a $4.5 trillion market for wireless technology over the next 10 years. But we can't have wi-fi hotspots everywhere, can we?

We can. Chamtech Enterprises has developed a metal can fill of a liquid which is full of nano-capacitors, which, when sprayed on a surface—any hard surface—can receive radio signals and transmit them more effectively and efficiently than a metal rod or antenna can. Add in a router, and the antennas can communicate with a fiber network or receive signals from satellites, just like a phone. Result: a "mesh" network of low cost broadcast wi-fi virtually everywhere. And no ugly cell towers.

SUBLMINAL MESSAGE GLASSES

The story goes (even if it's probably an urban legend) that in the 1950s, a movie theater in New Jersey flashed the words "Eat popcorn" on the screen, too quickly to be consciously perceived. Result: a massive increase in popcorn sales. Ever since, the idea of "subliminal messages" has been a curiosity for both the marketing and self-improvement industries.

In 1992 Canadians Faye Tanefsky and Michael R. McCaughey patented Subliminal Message Goggles in order to "impart a subliminal message to the wearer." The patent credits the stereoscopic effect of human vision to the glasses' ability to create a single image in the human brain when projected before both eyes. While it's not entirely clear how stereoscopic vision relates to the effect of subliminal messaging, it sure makes the description sound science-y.

The patent holders included an illustration of aviator glasses (the patent does not make it clear whether a mirrored version would be available), along with four helpful illustrations of the kinds of messages supported by the glasses. One is a cigarette with a *Ghostbusters*-esque "No" circle; another was the same with a glass of booze. There is also a smiley face and what appears to be a ping-pong paddle, so as to, we can only speculate, subliminally influence the wearer to enjoy ping-pong more.

THE WORLD'S WEIRDEST VENDING MACHINES

Suntan lotion. In 1949 the Star Manufacturing Company debuted this unusual invention at a vending machine convention in Chicago. It resembled a gas pump and gave users a 30-second "spray job" of suntain oil. Cost: 10 cents. The designers hoped the machines would become commonplace at public swimming pools, beaches, parks, or wherever people were too lazy to apply sunblock with their own hands, but only a handful were ever sold.

Bait. In 1993 Pennsylvania fisherman Joe Meyer converted an old refrigerated sandwich-dispensing machine into an automated nightcrawler dispenser and sold it to Bob Williams, a bait shop owner. Williams now owns a company that distributes the machines, which dispense more than 10,000 bits of bait a week–each–during the height of fishing season. Imitators have since cropped up; over a thousand bait machines can be found across all 50 states.

Whiskey. Evva debuted at a 1960 beverage exhibition in London: a bottled mix of whiskey and club soda, served cold. Just like unattended cigarette machines in restaurants and gas stations, Evva would have likely proven popular among teenagers, which is probably why the machines couldn't find wide distribution.

Curling irons. The U.K.-based Beautiful Vending created a vending machine in 2005 that can eliminate a bad hair day (or night) in just a few short minutes. Vending Stylers are wall-mounted contraptions that offer hair straighteners and curling irons. They typically cost a dollar for every minute of use. The machines can now be found in shopping malls, clubs, and fitness centers in over 35 countries.

Mashed potatoes. These amazing machines can't be found in stateside 7-Elevens, but they're immensely popular at the chain's locations in Singapore. For about a dollar, locals can, as the ad copy promises, "get their gravy on," literally, with a heaping helping of mashed potatoes and brown gravy. The machines, which look like your average Slurpee dispenser, were created by Maggi, a sauce and instant-soup company owned by Nestlé. (For those who don't like conventional mashed potatoes and gravy, barbecue sauce is also available.)

Cupcakes. Sprinkles, a Beverly Hills–based bakery chain, opened its first "Cupcake ATM" in March 2012. The pastel-colored machine cranks out up to 1,000 treats a day and has been known to crash due to high demand. It takes credit cards, offers over a dozen varieties ranging from gluten-free to red velvet, and dispenses the cakes in cute little boxes. It also plays a delightful song about cupcakes.

SPRAY-ON CONDOMS

If there's any product in the world that consumers need to absolutely trust, a product for which nothing can be off, for which mistakes are inexcusable, it's got to be the condom, that inexpensive piece of rubber millions use to prevent them from becoming parents or contracting a disease.

Which is why German inventor Jan Vinzenz Krause probably faces an uphill battle to get his invention, the spray-on condom, to penetrate the billion-dollar prophylactic market. Negating the argument that condoms diminish feeling, Krause simply found a way to liquify latex, and put it in a can. When the moment is right, a fellow doesn't simply spray on a coat of latex; Krause's "Jolly Joe" is comprised of a plastic tube into which a man inserts himself, which then sprays liquid latex from all directions. (It then takes a potentially mood-killing three minutes to dry.)

Krause, a sexual health educator, was inspired by "MacGyver," the machines at a drive-through car wash, and his inability to find a condom that fit properly (sure, dude).

THE KISSING SHIELD

It used to be that if you wanted to demonstrate your affection for someone but were paralyzed by the fear of being exposed to insidious mouth-borne pathogens, a hearty handshake was your best bet. But thanks to the inventor of the Kissing Shield, sanitary romance is not dead.

The shield is a thin latex membrane stretched over a heart-shaped frame (you know, for romance), and it even comes equipped with a convenient carrying handle. It's safe and practical, which of course are the cornerstones of any good romance. (Sorry, Romeo, you'll have to supply your own surgical gloves for holding hands.)

The inventor of the Kissing Shield states that it is intended for kissing the "recipient of the user's affection" (even the description is antiseptic) but is also useful for the old tradition of politicians kissing babies.

The natural evolution from the Kissing Shield is of course an entire line of hygienic romance products. We look forward to antibacterial-gel-filled candy, Clorox-laced Valentine's Day cards, and giant, heart-shaped plastic bubbles you can ride around in.

TOAST WITH THE MOST

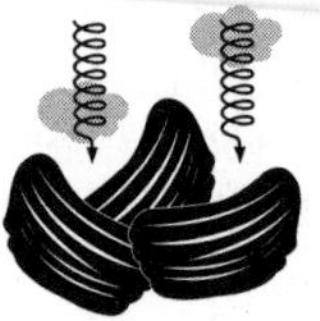

Reports of images appearing on toast are almost common these days—hardly anyone bats an eye when some guy in Ohio claims that an image of Liberace spontaneously appeared in the dark brown bits in his morning toast.

You no longer have to wait for the universe or your faulty toaster to bless you with an image burned into your bread, because industrial designer Sung Bae Chang has invented the Scan Toaster. It's simple, really—it's a standard toaster that connects to a computer via a USB cord. You use an imaging program to send a picture—any picture at all—to the toaster. The machine then etches the image into toasted bread.

This beats a similar item devised by a company called Yanko Design: the Note Toaster. Not much bigger than a slice of toast, this electronic breadbox features an e-ink display. The user writes a note on the display with a stylus—a grocery list or to-do list, for example—and the toaster burns it onto bread, creating an edible note to enjoy on your way out the door in the morning.

THONG DIAPER

If there's been one thing holding adult diapers back from widespread public acceptance, it's a notable lack of sex appeal. After all, how do you add some marketing oomph to a product whose sole job is to make sure your own oomph doesn't leak down your leg?

According to one enterprising inventor the answer is, of course, to replace the unsightly fasteners and excess material with thong straps. Because as Dame Fashion always says, "Maybe you can't control your bowels, but you sure as hell can control how fabulous you look!" While removing protective material from an adult diaper doesn't seem like the most prudent move, there is a tremendous upside. Thanks to the thong diaper's slimmer, less bulky profile, whether or not it successfully performs its primary function, you're sure to turn heads at the rec center.

Of course, the patent for the thong diaper doesn't explicitly state that it isn't actually meant for use with babies, but that alternative is simply too revolting to contemplate.

FACE-TO-FACE TANDEM BIKE

Chen Yugang took about a year to figure out exactly how to make it work, but in 2012 he unveiled his tandem bike…on which two riders face each other and somehow pedal in the same direction without smashing into the ground or their knees into each other. One of the riders does have to pedal backward, though, like in romantic slow dancing. That just adds to the atmosphere of what Chen thinks is a romantic bicycle.

Chen's device can be configured in a number of setups, based on the familiarity and relationship of the duo riding it. "Face-to-face is suitable for a parent and child, or dating couples," Chen says. The back-to-back mode gives each rider a good view (and not of each other). Standard, double-front-facing style can also be arranged, but where's the fun in that?

SQUATTY POTTY

Before flush toilets became commonplace, people did their business while squatting. While this could get messy, it offered health benefits that have been largely forgotten in the modern age. Squat advocates (a real constituency) argue that sitting at a 90-degree angle is not how nature intended us to poop. According to them, squatting can help prevent colon cancer, constipation, hemorrhoids, and "cardiovascular incidents," like the one that killed Elvis.

The Squatty Potty is a toilet adapter that lets users poop more easily from atop their regular toilets. The device fits over the seat and elevates the user at an angle more akin to squatting. It was invented in 2010 by Judy Edwards. A chronic constipation sufferer, she read about the squat method, and gave it a try. She stacked boxes and phone books in front of her toilet to serve as a squatting platform. This solved her problem, but having all that stuff sitting around was hardly convenient. After consulting with medical professionals to pinpoint the ideal height, position, and angle for squatting, Edwards and her son Robert created the first Squatty Potty.

More than 10,000 Squatty Potties have been sold. They're offered in three varieties: The "Classic" is sleek, white, and made of recycled wood; the "Ecco" is sturdy enough to be used by a large family; and the "Tao Bamboo" is both handmade and super-classy.

THE WORLD'S LIGHTEST SYNTHETIC MATERIAL

Answering the impossible question, "Why can't metal be lighter than a feather?" a quandary previously explored, in another context, by Poison and Skid Row during the 1980s, a team of California scientists created a spongy, ultralight "microlattice" from interlocking, hollow nickel tubes. Each tube is 1,000 times thinner than a human hair. When woven together, the substance is 100 times lighter than Styrofoam. It floats through the air like a feather and can rest comfortably atop a dandelion. Perhaps just as impressive, considering its metallic origins, the microlattice achieves a 98-percent bounce-back resilience when squished. (Unlike, say, the dandelion.)

A metallic material that can leave a dandelion's fluff undisturbed? There's got to be a *Horton Hears a Who* sequel in there somewhere. Then again, the researchers suggest that their microlattice might make a terrific acoustic-damping material for soundproofing walls. Another potential use? Impact protection for aircraft, from airplanes to spaceships, where lightness is fairly important. Here's hoping, however, that kids don't try using the stuff for trampolines or gymnastics mats; those little metal slivers can be murder when they get under the skin.

GENETICALLY-MODIFIED SILK

We all like to think we're Earth-friendly types—just maybe not friendly enough to voluntarily give up our dual-car households stuffed with plastic gadgets. Fortunately, while we've been sitting around feeling sort of guilty about our non-stop gorging on fossil fuels, scientists at the University of Wyoming have been busy working on a possible alternative.

It involves using cutting-edge science in tandem with a totally old-school tool; in this case, the humble, hard-working silkworm, who's getting a genetic facelift in order to produce silk that has the tensile strength of a spider's. As it turns out, scientists have long hoped to "farm" spiders in order to harvest their silk; weight-for-weight, it's stronger than steel, and presents all sorts of possibilities—medical implants, bionic ligaments, even tough, biodegradable plastic. Problem is, spiders don't produce much of it, and even if they did, their propensity for eating each other makes them difficult to keep in close quarters.

Silkworms, on the other hand, make plenty of silk (and they aren't cannibals), but it isn't as strong as the spidery stuff. Undaunted by their rude refusal to cooperate, the Wyoming University team simply stole spiders' DNA, and the result seems to be a genetically modified silkworm capable of producing super-strong silk—which is great and all, but also feels like the setup for a spider-based horror movie.

THE LEVITATIONARIUM!

Since the dawn of humanity, man has dreamed of flying like the birds in the sky (or at least that's what narrators always intone in aviation documentaries), but once airplanes successfully got off the ground, as it were, man's thoughts turned from flying to plummeting for some reason.

In order to help skydivers practice their craft a bit before actually leaping out of a plane, Canadian inventor Jean St. Germain honed existing wind tunnel technology to create the so-called Levitationarium in 1979, using propellers to produce an upward airflow within a chamber to effectively levitate those individuals within. St. Germain then sold his concept to the Aerodium Company, which has further perfected the Levitationarium. Sadly, it's now generally referred to by the far less snazzy name "recreational vertical wind tunnel," but, as Aerodium proudly notes on its website, one such tunnel could be seen during the closing ceremony of Torino 2006 Olympic Games as "the best bodyflyers from Latvia made spectacular show, flying at a height of 25 meters during the live broadcast to billions of viewers all over the world." Still, it would've been way cooler if they'd introduced them by saying, "Behold! The LEVITATIONARIUM!"

SILENT PARTIES

You know what the worst part about a dance party in which people get together to dance to loud, thumping, nonstop electronic music? All of the music of course. Who needs all that audible music delivered via speakers so as to be heard by all in attendance, right?

The concept of a "silent party" goes back to a 1969 Finnish science-fiction movie called *A Time of Roses*. Taking place in the distance, space-age future, partygoers are depicted dancing in a silent room. Brainwashed? No. They're all listening to the same music on individual pairs of headphones, sent out via wireless signal. That's a pretty obscure movie, but it may have inspired the "silent disco" trend that took off in the British isles in the 2000s.

In 2000 a BBC-sponsored "silent gig" was held in Cardiff, Wales. The hundreds of attendees listened to various DJs spin music through wireless headphones, to which the music was beamed. The concept is ideal for places with noise restrictions, or where curfews limit how late clubs or private parties may play loud music. If the place is silent, the music can go on well into the night. The only sound is the sound of feet shuffling.

SELF-DRIVING CARS

Along with food pills and video phones, the idea of self-driving cars has long been a staple of Western culture's collective idea of "the future," and thanks to the brain trust at Oxford University, those visions are finally on the verge of becoming reality. Led by Professor Paul Newman, a group of scientists has developed a self-driving system, which they've installed on a Nissan Leaf and have begun testing on university grounds. (Although the system still requires someone to be situated squarely in the driver's seat.) The programming, which uses 3-D laser scanning to build up a map of the vehicle's surroundings that's accurate to within a few centimeters, has been tested successfully at up to 40 mph.

Don't worry, there's no chance of the vehicle going all HAL 9000 on the driver: The system won't even offer to drive unless the road conditions match the necessary programming requirements. Newman even describes the scenario as "essentially an advanced driver system," sort of the next-next-next generation of power steering.

Although the current price of equipping a vehicle with the technology hovers in the range of $7,560, which is still pretty cheap all things considered. Newman's goal is to reduce the cost to a decidedly more reasonable $150.

THE MEDITATION BAG

People who meditate find that the hardest part of seeking serenity is cutting out all the immediate distractions of the outside world—ambient noise, discomfort as they sit on the ground, things like that.

Being able to turn off your senses so as to reach a higher mental or spiritual plane is the theory behind sensory deprivation tanks, in which a person floats in body-temperature water in the dark. Another way to cut out sensory input so as to look inward: the Sleeping and Meditation Bag, patented in 1982.

"Sleeping bag" is in the name, because that's its jumping-off point. It's a sleeping bag that you can wear while sitting up, in the cross-legged meditation position, or while lying down. It completely covers an entire adult and can be pulled tight with drawstrings to keep out the elements, distractions, light, and noise. And it's got a soft cushion to keep your tush comfy during those prolonged meditation sessions. On all sides is a thick layer of puffiness to send you on your way to spiritual bliss.

HIGH-FIVE SIMULATOR

Sometimes you need a bro to give you a high-five to celebrate a job well done, a witty barb, or other minorly awesome event in your life that necessitates extremely brief human contact. But what do you do if you're all alone? You get a mechanical High Five Simulator, bro.

Actually patented in 1994, the High Five Simulator is essentially a spring-loaded arm that's mounted on a wall, so it's always ready for a good slap. A fake hand attached to a forearm piece is connected to a lower arm section with an elbow joint for pivoting. When the hand is struck, the raised arm bends backward briefly before returning to the ready position.

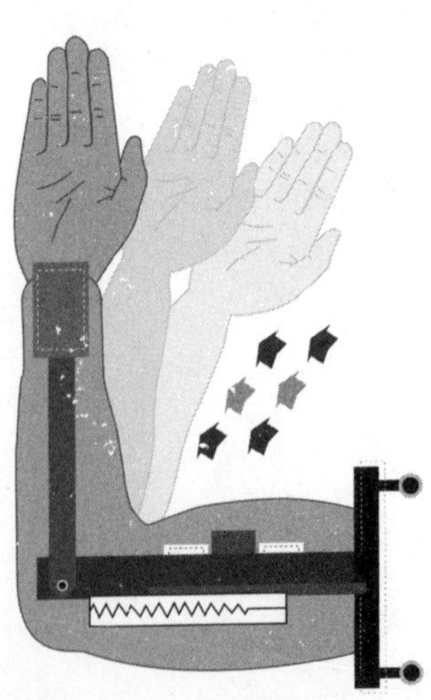

This invention is perfect for a lonely and excessive high-fiver, or just somebody who can't get their friends to satiate their need with real high-fives. One would think that this would be a perfect gift catalog item for dads, bros, and sports fans, but it never took off commercially. It did, however, show up on a 2013 episode of *30 Rock*, used by an editor of a men's magazine to high-five himself.

SHIRTS THAT ROCK

Electric-guitar shirt: The human desire to rock out is so profound that it often cannot be denied, but what if a yearning to kick out the jams kicks in when you're in no position to sling your six-string? Those who prefer to play it safe and maintain the option to riff at any given moment may wish to add the Electronic Guitar Shirt to their wardrobe. The black T-shirt features an artistic rendering of a white guitar–surrounded by orange, yellow, and red flames for added rock credibility–that can be played just like a real guitar, with each button on the neck triggering a different major chord. The shirt also comes with a mini amp with an adjustable tone knob that can be attached to your belt (which literally "goes to 11"). And the music-making circuitry is easily removable, making it easy to wash off all the rock 'n' roll sweat and groupie pheromones.

Graphic-equalizer shirt: Sure, having a playable electric-guitar shirt is cool, but you know what would make it cooler? Playing it while you've got a buddy standing beside you wearing an equalizer shirt (not a shirt advertising '80s TV series *The Equalizer*). The so-called T-Qualizer Raver, described on its website as "a sound-sensitive flashing equalizer T-shirt designed to heat things up while you move," features a thin, pliable electroluminescent panel that lights up and flashes to the beat of whatever music is playing in your vicinity.

NO MORE MISSING SOCKS!

For those who've been prematurely claiming that science has already invented everything that needs to be invented, you may at long last finally be right, now that the Swiss company Blacksocks has developed Smarter Socks. Expanding on their line of bestselling black calf socks, the company has begun to offer a Plus+ version of the socks that features an RFID chip that incontrovertibly identifies which socks belong to each other.

Mind you, this necessitates the purchase of the $189 Sock Sorter device that features the button that allows communication with the aforementioned chip (it does, it must be said, come with 10 pairs of the socks), but "the interaction between the socks with a communication button, the Sock Sorter, and an iPhone app makes sorting socks child's play," trumpets Blacksocks.com. As if it isn't enough that it helps you establish a record of how often your socks have been washed, the app also features a black-o-meter to determine the precise level of your black calf socks' blackness. Yes, really.

Of course, now that science has solved the mystery of where all the missing socks go, science will have to come up with a new go-to hack joke for stand-up comedians.

SQUARE WATERMELONS

Most of Japan's population lives in crowded metropolises like Tokyo and Kyoto, where apartments, while expensive, are often less than 800 square feet, scarcely larger than a dorm room. Japanese make do with small refrigerators, which certainly can't hold an unwieldy, oblong object like a watermelon.

In the early '80s, a farmer in Shikoku came up with an ingenious solution: He started growing watermelons in tempered-glass boxes. As the melons grew, their sides were impeded by the edges of the boxes. Instead of taking on a traditional, blimp-like shape, they became cubic. The square fruits, which are sometimes referred to as bonsai watermelons, are easy to transport, stack on market shelves, and store in tiny fridges.

They're also a luxury item. Only around 1,000 of them are grown each year, and they cost 10,000 Yen—around $114. They can't be found in an average Japanese grocery store, and are typically sold in high-end supermarkets and even department stores in posh areas like Tokyo's Ginza district.

While the square types are still popular in Japan, in recent years they've been upstaged by watermelons shaped like triangles, pyramids, and hearts, which are a trendy Christmas gift. Strangest of them all is a melon that's shaped like a human face. Cost of these more elaborate melons: the equivalent of $500 and up.

SMOKER'S HAT

Concerns about the dangers of secondhand smoke have led local governments to ban cigarettes in so many places that smokers can be forgiven for feeling they've been banished to Siberia. But back in 1989, one inventor created a contraption that might have fended off decades of municipal legislation: the Smoker's Hat. It was an apparatus that sought to nullify the health impacts—and the noxious odors—of cigarette smoke.

The concept was as ingenious as it was simple and fashion-forward, which is to say, not at all. The battery-powered, head-mounted device vacuumed up smoke that emerged from the user's cigarette, then sucked it through an ionizing and deodorizing filter. It even spritzed a refreshing scent before shooting the transformed smoke through an exhaust fan. For the smoker's convenience, the hat featured two cigarette-pack holders, a visor, and, to turn smoking into a dangerously hands-free activity, a clip that held the butt in front of the smoker's face.

THE MAN BRA

Finally,the male bra–previously known only as a sight gag called "the Bro" in an episode of *Seinfeld*–is a reality. Yes, Japanese men who want to enjoy the comfort and support of a bra need no longer skulk through the racks at Victoria's Secret (or the Japanese equivalent) and make up stories about trying on some delicate unmentionables "for my wife, who is not here right now." And for that you can thank WishRoom's Men's Premium Brassiere.

The confusing thing is figuring out exactly what group of men is supposed to find this appealing. All of the company's promotional materials show the bras (which don't seem particularly large in the cup area) being worn by chiseled, flat-chested mannequins. So it seems that it's not a good option for the overweight male sporting man-boobs, nor the fellow with overly developed pectorals.

That probably leaves cross-dressers looking for an undergarment that's both fun and practical. But then the question becomes, why not just buy the cheaper women's bras? Wouldn't that make more sense?

OLESTRA

In the late 1960s, Proctor & Gamble scientists were charged with creating a nutrition supplement to help premature babies gain weight quickly. They played around with sucrose molecules, manipulating them into different configurations in hopes that they would be more efficiently absorbed by the digestive tract. As is often the case, the science went horribly awry, and instead of creating something that would improve the lives of tiny, vulnerable babies struggling to survive, P&G accidentally invented Olestra, a substance that would allow people to eat as many potato chips as they wanted without feeling guilty.

Formally known as "sucrose polyester," Olestra is a synthetic fat substitute which is made by altering the chemical components of sugar and oil. When used as a food additive, it replicates both the delicious taste and satisfying mouthfeel of fat. However, the molecules are too large to be properly absorbed by the intestinal tract. The hoped-for result: food that tastes rich and fatty, but isn't absorbed by the human body and turned into body fat. A win-win, right?

Not exactly. The only problem, which turned out to be a significant one, was that if the fat you eat doesn't get absorbed into your intestines, it has nowhere to go but *out of* your intestines. In less delicate terms, foods made with Olestra can cause people to experience what the Food & Drug Administration describes as "abdominal cramping and loose stools." No one should have been surprised by this side effect; early safety testing with lab rats resulted in "anal leakage" and vitamin malabsorption. Because of this, P&G was engaged in a legal brouhaha with the FDA that stretched on for decades before Olestra was allowed on the market.

Ultimately, it was decided that Olestra could be used as a commercial fat substitute, with one caveat: Foods made with it had to carry a warning label so that customers would know not to stray too far from their bathrooms. In 1998, after 30 years of hard science and legal battles, Olestra made its debut in the form of Frito Lay's "Wow!" brand of snack chips. Sales were initially impressive ("Wow! A savory snack that won't make me gain weight!"), but then declined precipitously ("Wow! These fake chips are causing me to poop myself blind!").

Studies later indicated that Olestra's pants-ruining side effects were not as widespread as initially thought; i.e. not everyone who consumes the product suffers from uncontrollable diarrhea. That was good enough for the FDA, which no longer requires products made with Olestra to contain a warning label. Yet Olestra's bad reputation proved hard to shake, and to this day most people would rather gain a little weight than take their chances with it. Happily, science found another use for sucrose polyester, which has been repurposed as a machine lubricant and an additive in deck stains.

POTTY TRAINING FOR CATS

Apparently enough people have evaluated the budget spreadsheets comparing the kitty-litter costs with the water bill to warrant another member of the family getting in line for the bathroom. Hence CitiKitty Cat Toilet Training Kit, or the Original World Famous Litter Kwitter. Pet owners can wean their feline off the little sandbox in just a few steps: clipping a pseudo litter box under the toilet seat, then transitioning to a piece that has a big hole in the middle, and finally graduating to full-on toilet time!

Now lazy husbands have someone else to blame when they forget to flush. But your kitty won't be getting too many high fives from environmentalists, either; in 2007 the National Public Health Service for Wales wrote a scathing letter to the *Veterinary Record* citing cats as the source of the *Toxoplasma gondii* bacteria–typically found in cat feces–that were found in blood samples taken from numerous carcasses of whales, porpoises, and otters off the coast of England. A study by Swansea University soon after connected the bacteria to numerous people in the community who admitted to flushing their cats' droppings down the crapper.

CAT WIGS

If there's one thing cats hate, it's being messed with. And if there's another thing cats hate, it's when you try to make them wear stuff. Completely ignoring those two facts, Kitty Wigs are a thing.

Kitty Wigs are exactly what they sound like—wigs for cats. They were purely cosmetic items, because unlike humans who wear wigs to cover up hair loss, cats aren't hairless (except for hairless cats). But good luck getting a haired (or hairless) cat to wear one of these long, super-straight-haired wigs (they remind one of Cher in the early '70s). Four colors were initially available: "pink passion" (for a punk cat), "bashful blond" (for the cat who likes Old Hollywood glamour), "silver fox" (for the distinguished cat), and "electric blue" (for the cat who wants to have blue hair).

Kitty Wigs, the company behind Kitty Wigs, stopped selling them directly in 2012, and now publishes books full of cats wearing wigs.

X-RAY SHOE FITTER

Despite seeing their nation drop two atomic bombs on Japan in 1945, mid-20th-century Americans thought radiation was really cool and very futuristic. In fact, consumers didn't mind being blasted with a few gamma rays when they went to the store to make sure the shoes they bought fit perfectly. Hey, shoes are expensive, and beauty hurts.

In the late 1940s and early '50s, the Adrian X-Ray Company out of Milwaukee made and sold 10,000 devices to shoe stores that allowed customers to see just how well their shoes fit–by X-raying their feet while they were wearing the shoes. The X-Ray Shoe Fitter allowed them to see inside the shoe (and also their bones), via a small window on the unit.

The machines were a popular novelty–especially among children. They lasted in thousands of shoe stores until the federal government banned them in 1970. For while the box where a patron rested their feet was lead-lined, neither the compartment, nor the viewing windows were sealed. Result: persistant radiation leakage.

SPIRAL SHOES

Julian Hakes is a British architect and bridge-design specialist who decided to go a bit more intimate: He designed a shoe. "I always wondered if a traditional shoe is only the way it is because of the materials that existed when it was designed. What would happen if you started again?" he said.

Hakes started again. And he ended up with the Mojito, a weird spiral-shaped shoe. At first sight, it doesn't look like something a person could walk in: There's a thick, curved squiggle supporting the heel of the foot, and one holding the ball of the foot, but the arch is left exposed as a curve of the spiral kicks back to serve as a high heel. The resulting shoe is reminiscent of a twist of lime as imagined by Lisa Frank after a weekend in a dungeon.

The shoe was a hit among fashion bloggers before production even began in earnest, and it has been available since 2012. Women who hate their feet but love looking crazy can pick up a pair for about $200.

HEAD-SHAPED PUMPKIN MOLD

Things shaped like human heads, if they are not human heads, are creepy and unsettling. In fact, so are human heads, if they are not attached to human bodies. Also creepy: Halloween, with its witches and ghosts and vine-covered pumpkins sitting in fields at night.

These universal truths were completely overlooked or ignored by Ohio inventor John Czeszcziczki, who in 1937 patented "Forming Configurations on Natural Growths." In short, it's a mold you place onto immature pumpkins so that as they grow and ripen, they take the shape of the mold. And Czeszcziczki's main mold idea was that of a human head. This would result in pre-carved jack-o'-lanterns—Czeszcziczki evidently had trouble carving pumpkins, and in his patent he notes that "considerable skill is required to produce a good likeness." But other people must have been happy with their pumpkin-carving skills because Czeszcziczki's molds never took off.

In 1989 another inventor applied for a patent for molds that would have allowed zucchinis and other squash to grow into the shape of just the human face. Those didn't go into wide production either, but the company that made them, Vegiforms, remains in business.

WATERLESS DISHWASHER

For the past few decades, children have looked blankly upon their parents when informed that there was once a time when people had to wash their dishes by hand. But we now stand on the cusp of an era when kids will find it uncanny that water was ever a part of the process. Designed by Halit Sancar, Gökçe Altun, Pinar Simsek and Nagihan Tuna, the DualWash Bipartite Dishwasher forgoes our old friend H_2O in favor of its decidedly cooler cousin, CO_2, pumping supercritical carbon dioxide into the cleaning chamber.

Because of its low surface tension, the liquid carbon dioxide covers the surface of the dishes, and when the CO_2 returns to its gaseous state, any residue previously left on the dishes is forced into the filter, which can be removed and cleaned after each use. Voilà—your dishes are clean, and they're sterilized to boot! Plus, if it hasn't occurred to you yet, the general dryness of the DualWash not only saves a significant amount of water, but it also saves space by serving quite effectively as a china storage cabinet.

THE SELF-CLEANING HOUSE

By 1952 a 37-year-old designer and professional builder named Frances Gabe, of Newberg, Oregon, had had enough of the "thankless, unending, and nerve-twangling bore" of housework. So she designed and built a house that cleaned itself.

The house is built of cinder blocks to avoid termites and other wood-burrowing insects, and each room is fitted with a ceiling-mounted cleaning, drying, heating, and cooling device. The interior walls, floors, and ceilings are coated with resin to make them waterproof. The furniture is made entirely from waterproof composites. There are no carpets. The beds are covered automatically with waterproof material that rolls out from the foot of each bed. Easily damaged objects are protected under glass.

At the push of a few buttons, soapy water jets out from the ceiling to power-wash the rooms like an automatic car wash. The same jets then rinse off the soap, and then a huge built-in blower dries everything. The floors are sloped slightly at the corners so that excess water can run into a drain. The sink, shower, toilet, and tub clean themselves, too. So do the bookshelves and fireplace. The clothes closet serves as a washer and dryer, and the kitchen cabinets are dishwashers.

Gabe's been living in her patented house ever since–she's nearly 100 now, so maybe not doing all that housework has been good for her health.

VIDEO GAMES, WITH SMELLS

For the better part of a century, Hollywood has tried and failed to win, in addition to the hearts and minds of film patrons, their nostrils, repeatedly trying to introduce smell into the movie-going experience. For example: Smell-O-Vision, Odorama, and Aroma-Scope.

Either the folks at Scent Sciences and Sensory Acumen haven't read up on these past novelties/debacles in which smells were piped into movie theaters, or they must feel like they really know the noses of video-game players, because both companies have introduced systems designed to deliver game-appropriate odors in hopes of further immersing participants in whatever virtual reality they may be experiencing. Scent Sciences promises, "When synched with the action, scent-enabled media directs the release of ScentScape scents, providing background scents such as pine forest or ocean and from more scene related scents such as flowers and smoke to games or entertainment media." Indeed, in the description of the awesomely named GameSkunk, Sensory Acumen giddily warns potential buyers, "Just wait; you will be catapulted into a whole new way of game play."

CRIMINAL TRUTH EXTRACTOR

Fear is a great way to get people to let their guard down and tell you things they might not tell you when they have use of all their faculties and self-control. Also, what's the scariest thing in the world? According to this invention from the 1930s, a skeleton ghost.

The Criminal Truth Extractor was designed for police to use to get suspects under interrogation to admit to the crimes for which they were being questioned. Here's how it worked: The suspect sat in a chair in a small room. An officer in another room asked questions while viewing the suspect through a small window. (The whole thing would be recorded, too.) Also, the wall between the two rooms merely *looked* like a regular wall–however–it was extra thin, and when it looked like no good information was being given up, the interrogator could fire up the Truth Extractor–the lights went out and the thin wall lit up with he image of a ghostly skeleton with rapidly blinking eyes. The criminal gets scared, then confesses everything, so it went.

FLESH BRUSHING APPARATUS

Sure, it's important to flesh-brush everyday, but it's just so tedious and time-consuming. What? You don't flesh-brush? Well, it's no longer the 1880s, when the hottest hygiene fad going for the upper classes was using a soft brush to carefully scrub the body from head to toe and everywhere in between. It was so popular that in 1882 inventors Mary Stetson and William Bedell patented a mechanical Flesh Brushing Apparatus.

Saving the filthy filthy rich (or at least their servants) from having to scrub their bodies clean by hand, the Flesh Brushing Apparatus was a large, coffin-like tube that covered the body from the neck down and was lined with a dense forest of bristles made from "sea roots," whatever those are. A person simply climbed in, fully nude, and cranked a handle that was nestled amidst the thousands of brush fibers. This rotated the bristles around the body, which, according to the patent "effected a great saving of time and of exertion."

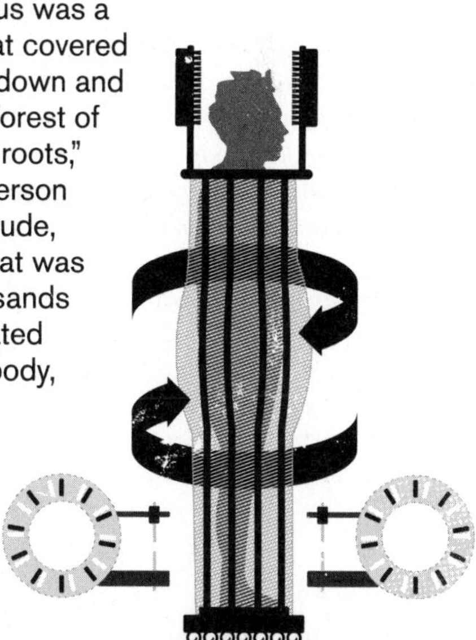

STAMP MOISTENER

In the dark days before self-adhesive stamps, people actually had to lick their own postage before affixing it to an envelope. It was positively barbaric; not only was it time-consuming and a waste of precious saliva, it tasted nasty too. We'd rather brush our teeth and drink orange juice right after. And don't even get us started on licking envelopes, what with the ever-present risk of debilitating paper cuts.

Thankfully, a creative fellow named Donald Poynter sought to deliver us from our long national nightmare with an invention cleverly described as a "device for moistening the adhesive coating on stamps and envelopes." Poynter's device is shaped like a box of tissues and has a plunger on top. When the plunger is depressed, a moistening arm shaped like human tongue moves up from a reservoir of water and sticks out of the box through an opening designed to look like a pair of lips. No explanation is provided for how this is more efficient than just using a damp sponge, or why someone would want to have such a creepy-looking thing on their desk.

THE VAPORTINI

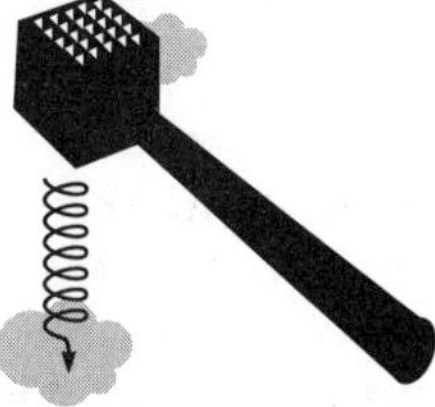

This product, introduced in 2013, looks like a glass orb with a straw sticking out of it—not unlike a crack pipe. But it's used to ingest another drug: alcohol. The Vaportini vaporizes alcoholic beverages so that users can get wasted just by breathing. The Vaportini works by way of a tealight that heats up your tipple of choice (the manufacturer recommends liquor of at least 70 proof, such as absinthe), causing it to evaporate into a fine mist that can be inhaled through a straw. All that pesky sipping, tasting, and savoring are over. For just $35, you can bypass your buzzkill of a digestive tract and absorb alcohol straight into your bloodstream.

The makers of the Vaportini claim that vaporizing alcohol removes all calories and carbohydrates, so you'll never have to worry about feeling full or gaining weight. You can just keep inhaling and inhaling to your heart's content, until your heart no longer works because you have died from alcohol poisoning.

TV HAT

Watching the two-minute commercial for TV Hat—not *The* TV Hat, mind you, just "TV Hat"—you start to feel bad for the folks who invented it. They obviously thought it was a million-dollar idea—a wearable shield that lets you watch portable video wherever you go—but failed to take into account their creation's fatal flaw: It looks darn goofy.

TV Hat is the quintessential late-night infomercial impulse buy, a product that seems like a good idea to customers in an altered state of mind at 2:00 a.m. The design is simply a ballcap with a flap of black fabric sewn along its freakishly long brim, forming a sort of rectangular tent that covers the user's face. Slip a smartphone or iPod into the elastic pocket at the far end, put in earbuds, and you're ready to watch video in complete privacy, with no glare, while keeping your hands free...so long as you don't mind looking like the Elephant Man wearing a World War I gas mask. It's like being in a private movie theater, where you cannot see or hear the inevitable mocking passers-by; TV Hat provides both a means to hide your face, and a reason to want to do so.

SMITTENS

Valentine's Day falls in February, of course. The problem: That's right in the middle of winter. Winter is cold in a great deal of the civilized world, and right around Valentine's Day, people are left with a choice: hold hands with their sweetie, or wear mittens. Of course, you could wear mittens while you hold hands, but it's almost impossible to get two hands covered in bulky fabric to stay locked together.

Clothing designer Wendy Feller came up with a whimsical idea to allow lovers to hold hands in the cold: Smittens (smitten + mittens = smittens). It's basically a pair of mittens sewn to each other to make one large mitten with two wrist cuffs and one large pocket. Each lover puts their hand in their side of the Smitten, and they can hold hands inside, protected from the cold. Smittens are made of warm polar fleece, and they have been a huge hit in novelty catalogs.

They come in black, blue, or red, and it's no accident that a Smitten looks just like a Valentine heart.

POLAR BEAR HATES SNORING

Snoring and sleep apnea cost thousands of Americans millions of precious zzz's every single night. A multitude of devices are available to aid the sleepless, but none of them is as innovative, or as strange, as the Jukusui-kun pillow.

This bizarre robotic pillow, which debuted at the 2011 International Robot Exhibition in Tokyo, could save your marriage. Users tuck themselves under a sheet lined with sensors that are attached to a large pillow shaped like a polar bear. When they start snoring or experience difficulty breathing, microphones in the bear trigger a robotic paw that tickles their forehead. This will, supposedly, encourage users to roll over and start breathing more normally without waking up.

As if all of this weren't cute (and weird) enough, the Jukusui-kun comes with a wireless monitor embedded in the tummy of an adorable teddy bear that attaches to the user's hand. It tracks blood-oxygen levels and can also activate the paw. While it sounds completely ridiculous, one of these things could help untold numbers of people get a good night's rest.

WILD WEST MOUSETRAP

The Old West was a dangerous place where everyday situations could suddenly take a turn for the worse. A game of cards may turn sour, a bank's customer might stick up the place, or someone might make an unneighborly visit to your land claim. No surprise, then, that it was the heyday of the Colt Single Action Army Revolver, the Frontier Six-Shooter, and other revolvers known by such specific nicknames as "the Civilian," "the Sheriff's Model," and "the Storekeeper." Less known: "the Rodent Controller."

In 1882, the same year that Robert Ford shot Jesse James, a wily gunsmith patented the Wild West Mouse Trap. Imagine a standard, spring-loaded mousetrap, only with a gun attached. It consisted of a wooden stand that held a trusty six-shooter at a downward angle, locked and loaded, pointed at a metal plate where bait would be placed. A metal arm ran from the front of the contraption to the trigger of the gun so that any varmints that tried to make off with the vittles would trip the switch. Rodents that weren't quick enough on the draw would be sent on to meet their maker. Like most mousetraps, this trap could only fire once before needing to be re-cocked and re-loaded.

There is no telling why this invention didn't catch on, although maybe it was because the idea of leaving cocked and loaded gun around might have been a bit too wild, even for 1882.

FUTURISTIC JAPANESE TOILETS

There's a good chance that you're reading this book while sitting on a toilet. A boring, uninspired, regular ol' toilet that, at best, is one of the newer "low-flow" models. But if you were reading this book in Japan, there's a good chance that you'd be doing your business on a more cutting-edge commode. Why Japan? It's a culture that loves both creative sanitation (sophisticated drainage systems were in place in the 8th century) and advanced technology.

While perfunctory squat toilets can still be found in public parks, much more sophisticated ones are commonplace in homes and offices across Japan. Typically referred to as washlets, these toilets offer a variety of nifty bells and whistles that are downright futuristic by American standards. The first high-tech toilets debuted in Japan in the early '80s and, after initial public skepticism, they soared in popularity.

Washlets often include features that are operated by video-game style controllers mounted on side panels. Most common among them is a bidet function that offers a wipe-free experience. Users can set the water temperature and control the pressure and direction of the stream. Other features include: heated seats, retractable cleaning wands, water massagers, air dryers, and, our favorite, automatic deodorizers. The seats on many washlets are automated and can be opened or closed with the push of a button or via an electronic motion detector.

Toto is Japan's best-known toilet manufacturer, and the company's "Washlet Zoe" earned the title "World's Most Sophisticated Toilet" in the 1997 edition of *Guinness World Records*. The Toto Neorest snagged another Guinness title in 2011 for "Most Functions in a Toilet." How many functions? Ten: automatic lift of seat and lid, auto rinse, energy saving, deodorant, a seat sensor, a heated seat, auto cleaning, washer for the user, air dryer for the user, and remote control.

In 2002 Matsushita created a toilet seat that can take a digital measurement of a user's buttocks and determine their body-fat percentage. Inax, another competitor, released a toilet around the same time that glows in the dark and helps people relax by playing one of six recordings of chirping birds, bubbling brooks, wind chimes, or Japanese harps. And many women's restrooms and private potties now feature a device called "The Sound Princess" that masks the noise of urination (a lot of Japanese gals find the sounds of nature completely mortifying).

Newer washlets offer air-conditioning for the summer and heaters for the winter months. Another one can measure blood-sugar levels via a device attached to a retractable arm. Future models may be controllable via voice commands and could even transmit health data to doctors over the Internet.

PERSONAL FIRE ESCAPE

While fleeing a building during a fire, the difference between life and death could be seconds—you've got to get out of there, and get out of there fast.

Benjamin Oppenheimer's 1879 invention did not get a person out of a building quickly, easily, or possibly even safely. His fire escape consisted of a pair of giant, shock-absorbent overshoes with thick elastic soles, plus an awning or parachute that attached to the head with a thick wire helmet.

In case of fire, you'd have to strap on the shoes and carefully place the helmet on your head, tighten the screws, and secure the chinstrap. Then, you could safely jump out the window and glide down to safety, with the parachute helping the journey to the ground and the shoes absorbing the impact of your landing. Just make sure the parachute doesn't catch fire.

LICENSE PLATE FLIPPER

Do you go off-roading or engage in some other activity that makes your car preternaturally dirty, particularly the rear license plate? Do you feel distressed because you are a good, law-abiding citizen who desires to keep your vehicle's identification clean and visible at all times, and wish to keep your license plate away from dirt and grime so that it's clean and legible when back on regulated city roads?

Well, then do you routinely drive way, way too fast and don't want to get caught by the traffic police?

Either way, the License Plate Flipper is a must-have. It's wired into your car's electrical system, and with a simple button push, it flips over the rear license plate to reveal, in its place, a previously concealed phony license plate. It takes just 1.3 seconds for the flipover to take place, which is about the same as the reaction time for a cop hiding behind a billboard to spring into action and try to catch up with you when you're going 90 in a 55 zone.

Cost of the License Plate Flipper: $445, which is more or less the price of a speeding ticket for going 90 in a 55 zone.

SAUNA PANTS

There are "hot pants" and then there are *really* hot pants, and it's most definitely the latter category that this particular pair of trousers falls into. Developed to provide all of the benefits of sitting in a sauna without wasting valuable time by getting any semblance of relaxation out of the process, the eye-poppingly orange Sauna Pants are trumpeted on AsSeenOnTV.com as possessing the ability to "make you sweat quickly in the areas where you need it most," meaning those troublesome fatty parts, such as the abdomen, waist, back, and hips. Like a real sauna, Sauna Pants help a person shed water and, technically, lose weight. Other assurances include the easing of tight muscles and sore joints, with a recommended usage time of a trifling 50 minutes per day.

Although the current model features an adjustable temperature control with four-inch cord, the premise of Sauna Pants extends back to a bygone era when electricity wasn't necessity. Indeed, an undated advertisement for Wonder Sauna Long Hot Pants that's made the rounds on the internet seems to promote an inflatable, hot-water-filled version of the pants—one endorsed by the USA's Amateur Athletic Union, no less—that's aimed at helping "health-watchers of America look better, feel better, [and] wake up your body" by purportedly enabling them to "slenderize exactly where you want."

THE ICE CREAM CONE ZONE

Many a modern-day invention is just a piece of molded plastic. It's how that molded plastic is shaped, and then put to use, that can change how things have been done for decades, solve problems we did or didn't know we needed to solve, and make millions for their inventors. Such is the case with the Buddy System. In 1997 inventor Bob Sotile ordered an ice cream cone at an ice cream shop and was grossed out when the worker handled his dirty money with her bare bands, then touched her hair, then handed him his waffle cone. So Sotile came up with the Buddy System (nicknamed "the Conedom"). Resembling a tiny, white traffic cone, it's used to grab a waffle cone, preventing any touching of the food. Bonus: As the customer eats the ice cream cone, the recessed top of the Conedom serves as a drip guard for melting ice cream.

Another ice cream innovation: the Motorized Ice Cream Cone. Should you tire of moving your mouth all around an ice cream cone, or even worse, turning the cone now and then to lick up drips or evenly consume the ice cream, this contraption does the work for you. Just place a scoop of ice cream in the cone-shaped plastic machine, press the button, and you can swirl your chocolate-vanilla with no muscle effort.

FINGER MUSTACHE TATTOOS

Tattoos have become completely mainstream, with college students, soccer moms, and businessmen deciding to get pictures permanently etched on their skin, joining the traditional demographics of bikers, prisoners, and sailors. Still, it's a big commitment to get a tattoo; but depending on where you get it, you can cover it up–lower back, calf, shoulder, for example, can all be covered up with clothes when one grows too embarrassed or old to pull off that cobra in a Yankees hat riding a Segway.

Harder to cover up: a tattoo on your finger, especially one that was only ever undertaken because it was sort of funny. In a trend that's taken off around Brooklyn, New York, college-age men and women get a permanent tattoo of a tiny handlebar mustache on one side of their index finger. Why? When they hold it up to their face, above the lip, it looks like they have a tiny, silly mustache.

The one drawback to a finger mustache tattoo (other than actually having a finger mustache tattoo) is that the joke doesn't work if you're wearing gloves. Problem solved: You can now buy gloves printed with a mustache on one side of the index fingers, which seems like a better idea, long-term.

SWISS ARMY RING

Who among us hasn't been in the awkward position of being trapped under an extremely tiny tree, with no tools at hand to free ourselves from almost-certain discomfort? Or perhaps we've needed to comb a stray moustache hair before an important client meeting, with no idea how to do it?

Well, if you wake up in a cold sweat dreading these, and only these, nightmare scenarios, you can now sleep easy, thanks to the Titanium Utility Ring. The ring, which is just 9 mm wide, is a multi-tool marvel that's sure to make normal-size tools obsolete. Folded inside the Titanium Utility Ring are the essential apparatuses for persevering in any situation: comb, straight blade, bottle opener, saw, and serrated blade.

With so many miniature tools at your disposal, the possibilities are endless. Why, you could open loosely capped bottles, dig your way out of a poorly constructed prison, or have the most immaculately groomed hamster of all your friends. In fact, the only item the Titanium Utility Ring doesn't come with is a sundial to measure how long it takes to do all of those things.

THE CIGARETTE RING

For many, the toughest part about smoking isn't the smell, the bad breath, the social stigma, or the health risks, but the manual dexterity involved in holding the cigarette between your fingers. In 1936 one clever soul, Watson P. Aull of St. Louis, tried to give comfort to clumsy smokers everywhere when he created the Cigarette Ring.

Worn on the index finger, the Cigarette Ring was a tiny clamp that sat on a base. Smokers would place the cigarette into the clamp, whereupon it could be puffed with ease, particularly by those who like to gesture with their hands when they talk. When finished, the smoker would simply unlock the clamp and dispose of the cigarette. The ring portion wrapped itself around the finger via a series of beads with a spring on one side. This allowed smokers of all shapes and sizes to pass theirs down from generation to generation (early, because of emphysema) without having to take them to the local jeweler to get them resized.

But although the ring angled the cigarette to keep the other fingers away from the lit end, Aull forgot to include a receptacle, meaning that the falling ashes would land directly on the user's hand.

MANNEQUIN GUITAR

No matter how weird an item may be, it's still a fair bet that someone somewhere will see it and say, "I must have it! I simply must!" Sometimes, though, you see something and find yourself a little concerned about what sort of person would get excited about an item like that, and the Mannequin Guitar is absolutely one such item.

Why? In short, because it's one of the creepiest things you're likely to ever see, let alone play. For reasons unknown, Lou Reimuller, a luthier by trade (that's someone who makes and repairs stringed instruments), decided that what the world needed was for someone to take a guitar and meld it into the abdomen of an armless mannequin of a little girl. "Sure, it freaks me out," you might say. "But how does it sound?" It's a valid question, but, frankly, we're not even sure that anyone beyond its creator has ever played the thing. Some might give Reimuller points for thinking outside the box, but most would likely take one look at this disconcerting amalgamation and call the authorities.

A WORKING MODEL OF THE EARTH

The problem with studying the Earth is that we're all standing on it. Imagine trying to describe the appearance of a house without ever leaving the interior. It can be done, but only with lots of inference and extrapolation, and with questionable accuracy.

Professor Dan Lathrop of the University of Maryland is tackling this problem head-on. He studies the Earth's magnetic field–the force that makes compass needles point north and helps shield the planet from solar radiation. The field is not a constant; its polarity flip-flops occasionally, and the field appears to be weakening overall with time. Lathrop wants to construct a theoretical model to predict the field's future behavior. His solution: Create a physical model, a mini-Earth to mimic the real thing on a manageable scale.

Lathrop's first attempt, a two-foot-diameter ball weighing 500 pounds, was unsuccessful. Since 2008 he's been working on the Three Meter Geodynamo–a 10-foot, 30 ton sphere. When filled with a core of molten sodium and set spinning at 90 mph, it will–in theory–provide invaluable data about the workings of our planet. And if this one doesn't work as expected, well, surely there are other uses for a gigantic model of the Earth. Perhaps a model-train layout.

WAR KITE

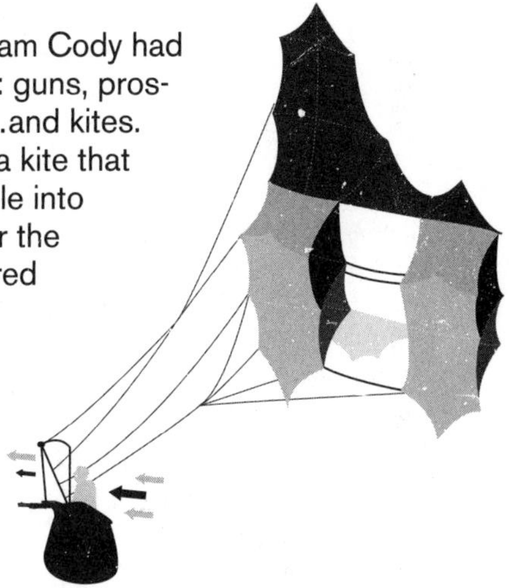

A cowboy named Sam Cody had severeal interests: guns, prospecting, horses …and kites. In 1902 Cody invented a kite that could lift a man half a mile into the air. During testing for the Cody Box Kite, he suffered numerous injuries, such as a broken arm and a near-drowning. Undeterred, Cody did what one does with any whimsical invention: He demonstrated his kite to the British navy so it could be used to kill people. The Cody Box Kite, in fighting mode, offered such features as a camera and a rifle, sort of like an extremely rudimentary fighter jet.

The navy declined to adopt the Box Kite for military use. Despite this setback, Cody went on to invent a variety of successful flying devices, including a plane that would become the first aircraft flown in British airspace. (Later, he managed to fly a plane directly into a cow, which is also neat.) He also created a Kite Boat, in which he somehow traversed the English Channel. In 1913 Cody died much as he lived: hurtling to the ground in a Waterplane, another of his inventions.

NEW WOOD

The production of wood products wreaks environmental damage, from habitat loss to pollution to splinters. Fortunately, progress is steadily being made towards sustainable alternative that will allow both trees and environmental advocates to breathe a sigh of relief. One possible option: Arboform.

This product could be the most exciting thing to happen to watery wood products since Wite-Out. Arboform was invented in the late '90s by a group of German scientists at the Fraunhofer Institute for Chemical Technology. They discovered that lignin, a key element of wood, can be transformed into a plastic-like material when mixed with resins and other natural substances. Unlike normal wood, Arboform doesn't require trees to be cut down, and it can be easily cast into just about any shape imaginable. It's also biodegradable and doesn't muck up the environment like common plastic products that require millions of barrels of oil to make every year.

Everything from baby toys to stereo speakers is now made out of Arboform, but TECNARO, the German company that produces the stuff, has yet to put the logging industry out of business.

THE PSYCHIC PLANT

It has long been accepted wisdom, if impossible to prove, that talking to one's plants can help them grow. An inventor took this a step further and found a way to make the plants talk back, and even possibly predict the future.

The Entertaining Growth System is basically a flowerpot that comes with a form-fitting lid that has two holes—labeled "yes" and "no"—cut out of it. Simply plant a seed as you would any other, then ask it a question. Whichever hole the seedling pokes itself out of is the answer, provided that the germination takes place before the question is forgotten. It's like a Magic 8-Ball, but with only two answers and no instant gratification.

Although phototropism means that the plant would poke out itself out of one of the holes, the inventor may not have realized that it would always go toward the hole that received the most light. Therefore, through proper placement, owners could literally bend the plants to their own will and game the fortune-telling processing a bit, the botanical equivalent of leaning on the Ouija Board gamepiece.

THE EYEBORG

When he was nine years old, Rob Spence lost his right eye in a shooting accident. Years later, the documentary filmmaker came up with an idea while looking at the camera on his smartphone. If a working, high-performance camera lens could be made small enough to fit into a handheld phone, surely something similar could fit inside his empty eye socket.

Spence called his friend Kosta Grammatis, an engineer. Together, they spent three weeks designing a robotic eye similar to the ones in the cranium of the unstoppable T-800 cyborg played by Arnold Schwarzenegger in the 1991 movie *Terminator 2*. Using a tiny camera donated by OmniVision, a company that specializes in pint-size photography equipment, they constructed a prosthetic device called "the Eyeborg." It fits inside Rob's empty socket. He also has a second version that glows red, much like the electric eye of a Terminator.

Unfortunately, the Eyeborg isn't connected to his brain and hasn't restored his vision. Instead, it shows others how Rob views his surrounding environment in real time via a handheld screen. Still, the prototype was listed as one of the "Best Inventions of 2009" by *Time*. While it's fairly rudimentary, the Eyeborg is a step in the right direction. With more advanced technology along these lines, maybe in another decade or two humanity will be able to say, "Hasta la vista, blindness." (Get it?)

INVISIBLE BIKE HELMETS

One of the drawbacks of commuting by bicycle (along with cars and the fact that it is exercise) is wearing a helmet. Many cities now require cyclists to wear them, but they totally mess up your hair and they're a pain to carry around—protection from head injury be damned.

But now there's a solution for those who cycle but hate to wear a helmet. In 2005 Swedish design students Anna Haupt and Terese Alstin designed the Hövding, or the "invisible bike helmet." The concept won them a grant, and after several years of tweaking, the Hövding hit the market in 2011. Not really a helmet, the Hövding falls somewhere between a collar and an airbag. Worn around the neck, it contains a protective bag that inflates in a fraction of a second if the wearer gets in an accident. The Hövding has a sensor that tracks "abnormal movements." If it detects something's amiss, a gas inflator fills the bag with helium.

The Hövding runs on a battery that's charged via a USB cord, much like a smartphone. The helmet also contains an airplane-style "black box" that records the acceleration and velocity of the wearer during an accident, information that could come in handy on an insurance claim. As if that weren't enough, users can order colorful removable liners so that their Hövding can match any outfit. A Hövding will set you back 499 euros, or about $665.

GREENHOUSE HELMET

Some of the world's biggest and best cities also have some of the world's biggest and best parks, perfect for jogging or bicycling. The problem is that those big cities are often horribly smog-choked, particularly on warm, sunny days; times that are preferred for outdoor exercise are rendered toxic by air pollution. The solution: the Greenhouse Helmet. More than just a plastic bubble protecting you and your lungs from the bad air outside, it's a complete mini-ecosystem inside.

It consists of a large plastic dome that seals firmly around the head. Inside are tiny shelves outfitted with plants, which, you might remember from junior-high science class, take in the carbon dioxide humans expel, and in turn expel oxygen. This creates an environment of a constant exchange of clean air.

And so you don't have to come across as a complete weirdo, the Greenhouse Helmet comes with speakers and a microphone for communicating with people who are not inside the Greenhouse Helmet.

REMOTE-CONTROLLED HORSE

There are few sensations that can compare to riding the range on a strong stallion, but if you're one of the few couch cowboys who can't be bothered to stop watching reruns of *Gunsmoke* and *Bonanza* long enough to hop into the saddle, fear not: In the early 1980s, an astute inventor came up with the perfect solution to your problem.

If the idea of a remote-controlled horse sounds laughable and self-defeating, that's only because it is. After equipping your trusty steed with the appropriate saddle, you'll be able to use your joystick to control the mechanized reins, thereby steering the horse in whichever direction you choose. For sadists, there's also the added bonus of being able to use the remote control to activate a whip to keep your horse in line. (This product has not been endorsed by PETA.)

Given how few jockeys have been replaced by this system, it's fair to say that this remote control has yet to revolutionize the world of horse racing, but with a few tweaks to New York City's budget, just imagine how relaxed those guys driving the carriages around Central Park could be.

INFINITE SOAP

"Waste not, want not" is one of those phrases you don't hear much anymore, but it's the kind of thing our Depression-era grandmothers liked to say. Which is why we wish they were still around to enjoy the Stack Infinite Soap Bar Cycle, a brilliantly goofy innovation that takes those words to a whole new level.

The idea here is pretty simple: Ordinary bars of soap, no matter how nice they smell or how good they are at cleaning us off, always end up as useless slivers that prevent us from getting every penny of our money's worth. To solve the problem, the folks at Stack devised a design with a "unique grooved shape" that you can use to "piggyback" your old soap onto a brand new bar, thus creating what they're calling the world's first "waste-free Infinite Cycle of Soap."

Of course, as our grandmother probably would have been quick to point out (while sternly wagging her finger), there's nothing stopping you from just squeezing your ordinary soap slivers into a plain old new bar that doesn't cost $13. But where would this great nation be if we ignored inventions that cost us extra money in the name of saving us money?

HUMAN SAIL

According to the sketches that accompanied the 1998 patent filing, the Human Sail looks like a full-body apparatus that transforms the wearer into a superhero, enabling flight and super-speed both in the air and on land. After all, it includes a strap-on backpack, mechanical wings, and inline skates. The truth is far less interesting: It's called a "body mounted sail assembly." Rather than turn a person into a super-human, it turns them into a human sailboat. The "wings" on the back are sails; the user flaps the wings, ever so gently, to capture the wind and propel themselves forward…on their roller skates. It's basic physics, nothing more, nothing less, and a great labor-saving device for someone who likes to roller skate but is too lazy to lean forward or propel themselves by their own power.

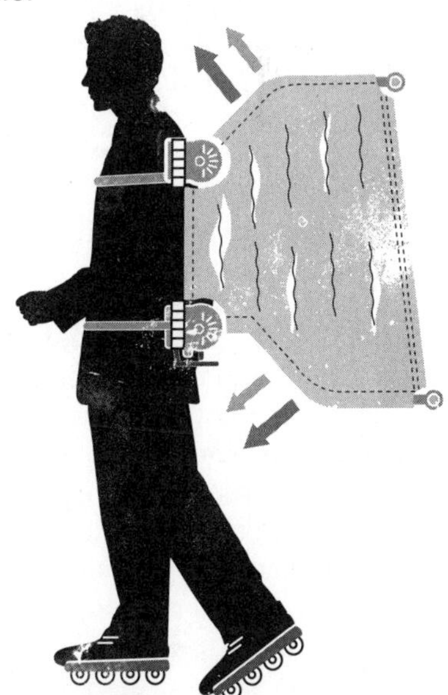

SADISTIC BABY GADGETS

The Baby Cage

Pediatricians have long debated the physical and emotional ramifications of "co-sleeping"—parents allowing baby to sleep in bed with them, rather than in a crib of their own down the hall. One genius sought to solve the very real and very frightening logistical concerns of co-sleeping (baby rolling off the bed, parents rolling onto baby) by inventing the Baby Cage, an oval dome that keeps baby locked inside and in place. And it's laced with curved crossbars sturdy enough to hold a parent's weight in case of a roll-over. So, hooray, baby survives the night...locked in a tiny prison.

The Diaper Alarm

One of the less appetizing perks of bringing up baby is the diaper check: the pat, the sniff, the grope that helps a parent figure out when it's time to make a change before the nappy runneth over. Putting a stop to such indignities was the goal of the Diaper Alarm, a battery-powered sensor that attached to the Huggies and initiated a mild electrical induction when wet, culminating in flashing lights and an audible alarm. A pool of pee-pee, a jolt of electricity, and baby's little privates—what could go wrong?

The Baby Patter

How far will new parents go for a good night's sleep? Apparently not far enough to make a fortune for the inventor of the Baby Patter, which provides a very loose interpretation of the word "soothe." Designed with the notion that an occasional tender touch keeps a sleeping infant content, the Patter attached to a crib and featured a motorized, robotlike arm—with "hand" attached—that sporadically would spring to life (gently, let's hope) and "pat" the sleeping baby's bottom. Can you feel the (robot) love tonight? So can baby...as long as it doesn't scooch around to face the opposite direction.

Cry No More

Millions of parents rely on the pacifier, that rubbery replacement for the comfort that only mom once could provide. Trouble is, babies eventually gain a pesky measure of autonomy, which too often results in the spit-out, the grab-and-fling, and other maneuvers that interrupt pacification and send parents diving under the sofa or reaching for the sanitizer. Cry No More was designed to end all that. It attached a binky semi-permanently to baby's face via circular straps that lassoed around the ears. Never mind that it left baby looking like Hannibal Lector in mid-transport and one bad breath away from choking.

LINCOLN TUNNEL CATWALK CARS

Buried 100 feet below the Hudson River, the Lincoln Tunnel is one of the world's busiest passageways, connecting New Jersey to Manhattan. On an average day, over 100,000 vehicles chug through its three tubes. And since 1961, officers working for the Port Authority of New York and New Jersey have had an invaluable weapon at their disposal to help fight crime and traffic jams tin the tunnel: Catwalk Cars.

When they debuted, the single-wheeled cars were fueled by gasoline as they sped along narrow catwalks attached to the edges of the tunnel's tubes. The 550-pound vehicles, which looked sort of like bumper cars and were constructed of aluminum and plastic, could move forward or backward at up to 30 mph.

Nowadays, officers monitor traffic in the Lincoln Tunnel 24 hours a day and are perpetually on alert. Given their proximity to New York City, the tubes are considered high-risk terrorist targets, but a more common threat to commuters is traffic delays. To help keep up with the ever-increasing number of motorists down there, the catwalk cars have been updated. The modern ones are lighter, faster, powered by electricity, and parked in 4' x 8' booths strategically stationed near the spots where incidents most often occur. Each of the tunnel's tubes has two of them. When things go wrong, an officer jumps into one of the tiny cars and races off like Batman to the rescue.

ONE INTENSIVE COMB-OVER

There are a lot of inventions out there claiming to cure baldness or regrow hair—helmets, lasers, creams, and surgeries, for example. But this invention is merely a technique for combing hair in a particular, yet very elaborate, way so that men who have lost all the hair on the tops of their heads (but not the sides) can make do with what they've got and trick any observers. (Also: You can indeed patent a technique or method of doing something with the U.S. patent office).

Registered in 1975, this "method of concealing partial baldness" takes the old trick of flapping a wave of hair from one side of the head over to the other to cover a bald spot and takes it to the next level. First, a man who is bald on top must grow out his hair in the back and on the sides, as long as it will go, until it hangs down like floppy drapes. Then, the method instructs, he carefully combs over each section, one at a time, layering them. The end result: short hair on the sides, full hair on top. (Keep it in place by spraying down each section with hair spray once it's lying flat.)

NEW FRONTIERS IN CAFFEINE DISPERSAL

Because drinking a cup or a can of something can take too long, you can now get these products, all of them infused with caffeine and other stimulants:

- Gummi Bears
- Maple syrup
- Breath mints
- Soap
- Gum
- Hot sauce
- Marshmallows
- Body wash
- Time-release caffeine capsules
- Dissolvable strips
- Beef jerky
- Popcorn
- Bloody Mary mix
- Breath spray
- Bottled water

DUMB USB GADGETS

Pet Rock: The item that became synonymous with the whimsical and gullible 1970s has returned! Only now, it's high-tech. In the spirit of the original, it does nothing; plugging it into your computer's USB port doesn't even draw any power away.

Squirming Tentacle: People will think an octopus or the fictional alien monster Cthulhu has taken control of your laptop when they see a moving tentacle coming out of a USB port. Unlike the Pet Rock, at least it moves.

Power Hour Album Shot Glass: This 1GB external storage drive contains 60 one-minute drinking songs. It fits perfectly into an included shot glass, used to play a game where the user takes a shot after every song.

Hot Cookie Cup Warmer: Working hard will never again mean that your long-ignored cup of coffee will go cold. This miniature hot plate, which resembles an oversize Oreo cookie, plugs into your computer and keeps a mug warm while you're focusing on a problem, attending a meeting, or trying to advance to the next level of *World of Warcraft*.

Fishquarium: Containing enough room for a couple of goldfish, the Fishquarium plugs into your computer, which powers a low-voltage pump and filtration system while the sounds of nature soothe away stress. It also has a pencil holder.

CORDLESS JUMP ROPE

Jumping rope has long been established as a superb cardiovascular exercise that helps get your whole body working hard, and it looks cool in training montages in boxing movies, but there are still some individuals whose crippling fear of rope and rope-related products prevents them from ever enjoying this leisure activity.

Or at least that's our presumption, anyway, as it seems like a slightly better excuse for the existence of the Cordless Jump Rope than someone just buying a piece of rope. According to the patent granted to Lester J. Clancy of Mansfield, Ohio, his invention features two handles, but instead of being attached to rope, "a donut-shaped enclosure is provided and mounted to the handle," inside of which is a weighted ball that, when rotated, will "generate rotational torque to simulate the use of a jump rope." While the Cordless Jump Rope does admittedly remove any possible chance of the user somehow accidentally getting rope burn, the value of which cannot be underestimated, it's hard not to consider the fact that it might prove more financially sound to bypass buying the product altogether and just, you know, jump.

WINE-BOT

Scientists at NEC System Technologies in Japan have invented a robot that can taste and identify dozens of wines, and even some foods. The green-and-white tabletop robot has a swiveling head, eyes, and a mouth that speaks in a child's voice. To identify a wine, the unopened bottle is placed in front of the robot's arm. An infrared beam scans the wine–through the bottle–and determines its chemical composition. The robot then names the wine, describes its taste, and recommends foods to pair it with.

Scientists are still working out the kinks: At a 2006 press conference, a reporter and a cameraman put their hands in front of the robot's infrared beam. According to the robot, the reporter tasted like bacon, and the cameraman tasted like ham.

COLLAPSIBLE RIDING COMPANION

There are a lot of reasons for not wanting to drive alone, or rather, for not wanting it to *look* like you're driving alone. Perhaps you have to drive home late, or through a sketchy neighborhood. Maybe you like to talk to yourself, or sing along to the radio, or argue with the pundits on talk radio, and you don't want to feel weird about it when you pass by other motorists. Whatever your reason, this 1991 patent filed by Rayma Rich of Las Vegas is the answer to your bizarrely specific needs: the Collapsible Riding Companion.

The CRC is essentially a mannequin head and torso outfitted with a full head of hair, a T-shirt, and a jacket (in case you like to run the A/C really high). Simply place it in the front seat next to you while you drive, and you've got a quiet but agreeable road-trip buddy.

In the event that you have a real person riding in the car with you, the Collapsible Riding Companion folds down and can be stored in its very own rectangular, suitcase-like carrying case.

BUBBLE HAT

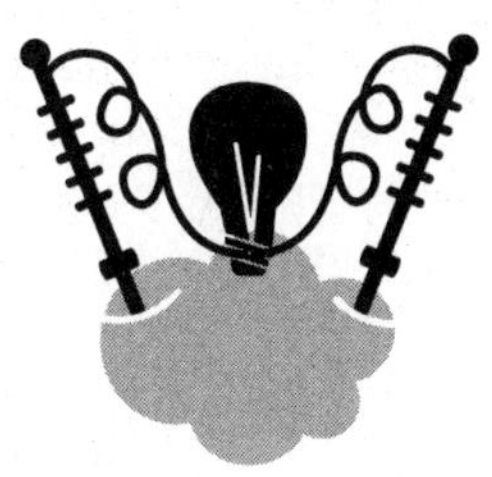

Elaborate ladies' hats have mostly fallen out of daily use, but in the early part of the 20th century, it was the height of fashion for a woman to wear a large, fancy hat. But if everybody was wearing one, how could a lady differentiate herself and draw attention to herself and her chapeau? With bubbles, that's how.

In 1912 Alden McMurtry invented the Bubble Hat. On first glance, it resembled any other piece of fine headwear outfitted with a large brim and false flowers. But hidden inside was a small chamber filled with soapy water. A tube ran from the chamber to a tiny tank filled with pressurized hydrogen that the woman would hold in her pocket or pocketbook. When the lady released a valve on the tank, it triggered the hydrogen to run into the soapy water, thus unleashing a torrent of delightful bubbles.

McMurtry thought the bubble-spewing device would be a great visual aid for choirs.

SPINNING WALL

Here's a radical solution to the problem of cramped apartment living: making use of the "dead" vertical space of a room. It's not a new idea (Murphy beds, room-splitting screens, e.g.), but the TurnOn Multi-Functional Spinning Wall takes it to extremes.

"Wall" is something of a misnomer. It's actually more of a wheel–an open-ended plastic cylinder with built-in furnishings extruded from its inner surface. The entire cylinder rotates to allow access to any of three functional areas of the module. After dining in the two-seat breakfast nook, residents can use their body weight to flip the unit over like a hamster wheel, rotating the dining table away to make room for a molded-plastic lounging couch. At day's end, another turn brings a flat sleeping platform into play.

A couple of these units, placed end-to-end, can theoretically create a usable living area within a tiny footprint. There are proposals for a food-prep module–with refrigerator, sink, and working stove–and a "Wet Cell," with toilet, shower, and vanity set at 120-degree radial angles; no mention of how the electrical and plumbing connections would work. Or of who, besides astronauts, submarine crews, or characters in a play, could stand to live like this.

CHOW, BELLA

Bella makes dozens of low-priced kitchen appliances, including coffee pots and slow cookers. But it's more notable for its single-purpose novelty bakeware that allows the user to make exactly one thing per device. Here are some Bella things you can't possibly live without (and probably got as a holiday gift last year):

- Circus Waffle Maker: turns out mini waffles shaped like elephants, giraffes, and monkeys
- Mini Donut Maker
- Mini Cupcake Maker
- Cake Pop & Donut Hole Maker (which are almost the same thing)
- Brownie Maker
- Ice Cream Sandwich Maker (cookie parts only)
- Cakesicle Maker (slightly larger cake pops)
- Pretzel Maker
- S'mores Maker
- Waffle Stick Maker (half of a waffle)
- Small Pot Pie Maker
- Pastry Tart Maker
- Strawberry Shortcake Maker

CONTROL-ALT-DELETE HANDLE

Computers break down. A lot. Especially if you've got an old one that doesn't work so good anymore and you have to shut programs down when that beach ball keeps spinning or that hourglass never disappears. So what do you do? You hit those three "shutdown instantly" buttons, all at the same time: control + alt + delete.

Pressing buttons? Three of them? All at the same time? That's, like, more work than an office worker is accustomed to or prepared for, and you have to use, like, two hands. If only there were some kind of tool to simplify this incredibly easy task! There's got to be a better way! There is: the Control+Alt+Delete Handle.

It's a T-shaped metal bar. One part is the handle, and the rest is a long metal bar with little pieces of rubber spaced out just right, so when you press down on the keyboard with one simple motion, all three buttons get pressed at the same time.

It literally takes less effort to press all three buttons at once than it does to reach even six inches to grab your Control+Alt+Delete Handle. This invention was something of a joke, concocted by a frustrated home-computer user. It spread around the Internet and was never mass-marketed, but still—somebody actually invented it.

THE I-ON-A-CO CURE-ALL BELT

In 1925 magazine publisher and health nut Gaylord Wilshire unveiled his invention, the I-ON-A-CO Electric Belt. Wilshire claimed that this seemingly revolutionary device could cure heart disease, all cancers, both types of diabetes, prostate issues, and any other illness he forgot to mention. How? Worn properly, the belt sent a mild jolt of electricity coursing through the body, which Wilshire said magnetized the iron in the blood, shocking the cells into behaving properly. "All you have to do," one ad read, "is to place over your shoulders the I-ON-A-CO. That's all. You may then light a cigarette and read your newspaper for 10 or 15 minutes."

Wilshire claimed that the belt could even cure medical ailments after a single application. This, and all the other claims, attracted the attention of the American Medical Association, the Public Health League, and the Better Business Bureau, among other watchdog organizations. After a scientific study, the AMA announced that the I-ON-A-CO was about as helpful in curing illness as "the left hind foot of a rabbit caught in a churchyard in the dark of the moon." Wilshire died just two years later. (The I-ON-A-CO belt evidently did not cure death.)

A SAFER COCKFIGHT

For most of us, cockfighting is a fringe phenomenon with a funny name, and the exclusive domain of degenerate gamblers in back alleys betting on battles between birds. The truth is, though, that it's one of the world's oldest spectator sports. And while it's illegal in all 50 states, underground cockfighting is still popular enough to draw the attention of law enforcement throughout the country. All of which is to say that we admit that the existence of a patented non-lethal cockfighting system is a very good thing.

In fact, this gadget—which swaps out the deadly knives known as "gaffs" that roosters wear in traditional cockfighting for an array of sensors that record the blows landed by an enemy bird—probably has more chicken-lifesaving potential than any other weird invention in this book. And given that it uses technology no more complicated than your average garage-door opener, it seems safe to assume it could be mass-produced relatively cheaply.

THE PLASTIC WISHBONE

Most weird and pointless inventions come about because of accidents, opportunism, or general greed, but one of the weirdest and most pointless inventions of them all was born of genuine strife and ennui. Ken Ahroni always looked forward to the tradition of breaking the wishbone after a Thanksgiving meal. The problem? There's only one wishbone per turkey, and it takes a mere two people to break it, so Ahroni often had to share the honor. One year, while seething over the injustice, his mind snapped like so many sad, brittle turkey clavicles. Unwilling, nay, *unable* to take it anymore, Ahroni started Lucky Break, a company devoted to the manufacture of plastic wishbones.

Lucky Break's artificial wishbones are exactly what they sound like: cheap little white pieces of plastic that look like turkey bones. You pay money for them, snap them in half, and then immediately throw them in the garbage. Now, we've already established that the inventor of such a product must be sort of a loon, but Ahroni is not to be trifled with: In 2008 he successfully sued Sears, Roebuck & Co. for stealing his idea. Big-timey lawyers for Sears hilariously argued that "any originality inherent in a replica of a wishbone was invested so by nature, by a supreme being, or by the turkey itself." In the ultimate lucky break, Ken won out to the tune of $1.7 million.

ACT NOW! SUPPLIES ARE LIMITED!

The best stuff ever advertised on late-night TV.

The Thighmaster. Sitcom star Suzanne Somers helped revive her sagging career by becoming a spokeswoman for this odd, butterfly-shaped exercise device. A series of infomercials featured the blonde bombshell using the ThighMaster and extolling the benefits of "squeezing your way to shapelier hips and thighs!" Late-night talk show hosts like David Letterman loved mocking the silly device, but Somers laughed all the way to the bank. Over 10 million ThighMasters have been sold to date.

The Rejuvenique Electrical Facial Mask. This one's definitely the weirdest product on our list. The Rejuvenique was a motorized mask that debuted in 1995 and was hawked on late-night TV by former *Dynasty* star Linda Evans. Powered by a 9-volt battery, it supposedly gave users a "facial workout" comparable to eight sit-ups, or "face-ups," as the ads called them. Those willing to wear the mask (which made them look like the love child of C-3PO and Hannibal Lecter) for 15-minute intervals, three to four times a week, were supposedly rewarded with a more youthful-looking face. It should come as no surprise that the Rejuvenique was the subject of an FDA investigation in the early 2000s.

The Flowbee. A San Diego carpenter created this "haircut vacuum attachment" in the late '80s and started selling them out of his garage, marketing it to families fed up with the high cost of professional haircuts. The rather ridiculous Flowbee caught on after appearing in a series of infomercials, but it's been widely panned by barbers and users alike. The problem: The device just isn't very good at cutting hair–it did only one style of haircut (and sucked up the cut hair), somewhere between a bowl cut and a pageboy, a look that is neither flattering nor hip. It became a national joke in the '90s, popping up in numerous TV shows and films like *Wayne's World*, which featured a parody version called the Suck Kut. (Inventor: "As you can see, it sucks as it cuts." Wayne: "It certainly does suck.") Nevertheless, over two million Flowbees were sold by 2000.

The Tiddy Bear. It took decades, but someone *finally* came up with a way to prevent "shoulder pain" caused by seat belts. While that's the alleged purpose of the Tiddy Bear, its name, and the voluptuous models featured in ads for the goofy product, suggest otherwise. The bear is, more or less, a Beanie Baby with a strap that fits over a seat belt, thus allowing for a more comfortable commute (if you happen to be a lady or a particularly buxom bloke). Talk show host Ellen DeGeneres once declared the Tiddy Bear "one of the best inventions I have ever found."

ANTI-GRAVITY FLYING SAUCER

The flying saucer: It's not just for aliens, by aliens! In the 1920s, American physicist Thomas Townsend Brown was studying the use of gravitational fields as a means of propulsion–in other words, using the power of gravity to make things go. He found that when he charged a capacitor to a high voltage, it moved toward its positive pole, creating an "ion wind." He claimed that this effect proved a link between electrical charge and gravitational mass, and argued that it could be harnessed to create flight, seemingly free of both the strictures of mainstream physics and the need for gasoline.

In 1953 Brown demonstrated his "electrogravitic" propulsion for the U.S. Army at Pearl Harbor by flying a pair of metal disks around a 50-foot course. Energized by 150,000 volts, the disks, which were three feet in diameter, reached speeds of several hundred miles per hour. According to Brown, the military immediately classified the project and no more was heard about it. But throughout the 1950s, Brown's work was cited as a possible explanation for how UFOs might be able to fly.

FANCY MAN MUSTACHE ACCESSORIES

The Mustache Cup and Glass. Patented by a man named Albert Schenck in 1879, this strange invention was designed to keep 'staches out of hot drinks. Picture a children's sippy cup crossbred with a coffee mug. In addition to a side chamber that allowed liquid to pass into the drinker's mouth, it also contained a miniature shelf on the top to keep facial hair out of the way.

The Mustache Guard. This one was created by Ruben P. Hollinshead in 1890 and was intended to help fanciful lads with mustaches eat meals. The guard, which sort of resembled a metallic bow tie, slipped over a gentleman's facial hair and was held in place with strings that fitted around the ears. While practical, and capable of keeping a 'stache from getting in the way of spoonfuls of delicious soup, it was probably incredibly uncomfortable to wear.

The Mustache Trainer. This harebrained device, invented by Louis Auguste Allard in 1889, looked like a cruel dental appliance. Supposedly, it allowed the wearer to "train" his mustache and make it grow in a "desired form and position" with the use of hooks. Needless to say, the trainer wasn't very good at helping men grow anything other than painful facial welts.

RELAXATION CAPSULE

In 1998, Dr. Claude Rossel opened his private clinic, Centre Biotonus Clinique Bon Port, with the intention of taking both the physical health and mental balance of patients into account. For those who can't afford to pop over to Switzerland at the drop of a hat, however, the clinic has designed the Relaxman Relaxation Capsule.

If you're stressed out about your stress levels, finding yourself increasingly desperate to unwind, and—we really can't stress this particular bit enough—you have $50,000 lying around, then you could do a lot worse than spending it on a Relaxman. Purportedly soundproof, lightproof, and heatproof, the capsule features a water mattress that remains heated to body temperature and plays pre-programmed music to soothe whatever savage beast lies down inside.

The advertisement for the Relaxman cites source-free research that conveniently shows that "a 50-minute rest in the negative ion-enriched atmosphere effectively helps reduce tension, anxiety, depression, and fatigue." The capsule is also supposedly pretty good for fixing folks' jet lag and sleep imbalance, as well it should be.

DISSOLVABLE MOUTH BURN STRIPS

Mouth burn. We've all had it happen. That slice of hot pizza just looks so good, you can't help but chomp down on it. And...*youch!* You burn the roof of your mouth. Or you're just too eager to chug down that cup of coffee, and your tonsils get singed. Fortunately, the mouth heals very quickly—usually within a day or two—but while it does, you have no choice but to suffer through the pain. Until now.

Jason McConville, an associate professor of pharmaceutical sciences at the University of New Mexico, has come up with an ingenious solution for mouth burn. His team of researchers have developed a strip, similar to the ones used for breath fresheners, that delivers benzocaine, a local anesthetic used by dentists and in cough drops. The strip is applied to the burned area inside the mouth and slowly dissolves, releasing the benzocaine and providing sweet relief from the pain. There's still a long approval process ahead, so it may be years before the strips appear in drug stores. But one day, we will at last have an over-the-counter remedy to a problem that has plagued humanity since the first intersection of cooking and impatience.

THE BIG HAIR HAT

This was patented in the early 1960s, when big hair was in fashion and big-haired ladies like Lady Bird Johnson were style icons. But Lady Bird was from Texas, where it doesn't rain all that much, so she and other large-coiffed Texas ladies didn't have to worry about the elements as much as the ladies around the country whose hair they were copying.

Sure, a hat would protect hair from rain and snow, but it would crush the delicate 'do. The Big Hair Hat guarded against the elements without putting undue pressure on the hair. It was an extra-thick, extra-tall, rigid shower cap. Simply slip it over the coif to protect the hair, while also looking like the pope.

SOUND PERFUME GLASSES

Have trouble remembering people after being introduced to them? A joint team of researchers in Japan and Singapore have developed a solution. Sound Perfume Glasses are high-tech specs that connect wirelessly to your smart phone. When you are introduced to someone, an app in your phone assigns your new friend an identifying sound as well as an associated scent chosen from among the eight solid perfumes stored in little pods on the earpieces of your glasses. Whenever you encounter that person again, your phone will connect to theirs, triggering your Sound Perfume Glasses to emit that person's associated sound from tiny hidden speakers, while heated wires in the earpieces activate their signature fragrance, thereby reinforcing their identity—and, the developers claim, promoting deeper, more pleasurable emotional bonding.

That sounds a little weird, but filmmakers have been associating characters with evocative sounds for decades. (Detective John Shaft would be less memorable without his iconic theme song, for example.) And research shows that scent is an extremely effective trigger of emotional memory.

But it takes only a single glance at someone's chunky white goggles with smells coming out of them to determine if this is a person you even want to know.

STYLUS ICONS

This has happened to you before: Your hands are just too full to drink your coffee and play with your touch-screen smartphone at the same time. What on Earth is a modern multitasker to do? The answer should be obvious: Attach a long stylus to your nose, so you can poke at your phone with it.

The nose-extending stylus was dreamed up by Dominic Wilcox, an artist who felt as though there simply weren't enough options for those who wanted to use their touch-screen devices while in the bathtub. While it's not commercially available, Wilcox did create a prototype. The long, cylindrical facemask nose stylus appears to attach to the head via two white shoelaces and comes straight from the 1999 film *Eyes Wide Shut.*

In other stylus news: For those of use whose chubby digits have been creating explorations of the dark recesses of AutoCorrect, there's now a fingertip stylus. The idea is this: You wear the stylus on the tip of your finger, sort of like a cool goth fingertip ring, then when the mood strikes, you've got a fingertip so dainty only a couple of angels could dance upon it.

TAPEWORM TRAP

There's got to be a better way to get rid of the common tapeworm, that parasite that lives in the human intestines and sucks away all of the nutrition you put into your body.

Okay, so it's not much of a problem anymore, at least not in the developed world, or in places where basic sanitation precludes the once-frequent passing of parasites from one person to another via exposure to feces. In the 19th century, though, it was a real problem, and because it was the 19th century, doctors were quite stymied as to how to remove a tapeworm without invasive, highly dangerous surgery.

A physician named Alpheus Myers invented a tapeworm-removal device in 1854 that didn't "employ medicines" or "cause much injury." That doesn't mean it was pleasant. Myers's gadget was a cross between a plumbing snake and a fishing pole. After fasting for a day or to make the worm hungry, the patient then swallowed the device, a three-inch-long metal trap on the end of a metal chain. The trap went into the stomach; the other end hung out of the person's mouth. The trap, outfitted with "any nutritious material," would lure the worm and grasp its head, at which point the patient would drag out the trap, and the worm along with it.

TALKING BASEBALL CARDS

By the late 1980s, kids (and adults) were buying baseball cards not just out of love of the game, but also as an investment, fooled by a marketing campaign to make them think that buying mass-produced pieces of cardboard at inflated prices would make them rich someday. In 1989 LJN Toys tried to cash in on the newfound love of the old pastime, but at least they aimed to make it about collecting and statistics again.

They came out with the Sportstalk–a handheld device about the size of a Walkman that "played" electronic baseball cards. Each card, which looked like a normal baseball card, only slightly larger and slightly thicker, had a tiny vinyl record embedded in the back. The Sportstalk then just played the record. Through the built-in speaker the size of a quarter came two minutes of statistics about the player (voiced by nine-time All-Star Joe Torre), along with radio calls of famous plays, and players reminiscing about their biggest moments on the field. It probably failed because it cost too much–$28 for the player and $2 per card. Toys "R" Us ordered half a million Sportstalks and sold fewer than 100,000.

VACUUM TRAIN

The train of the future may well be a VacTrain, a "magnetic levitation" train that, theoretically, will travel at extremely high speeds through vacuum tunnels. Engineers are currently looking at the VacTrain as the basis of a global subway network between continents and even under the oceans. The lack of air resistance in a vacuum tunnel would allow a VacTrain to reach speeds of more than 4,000 mph, or five to six times the speed of sound. The 3,100 mile trip from New York to London would take about an hour.

The concept of intercontinental tunnel travel is not new. Robert Goddard, the father of American rocketry, was issued 2 of his 214 patents for work on VacTrain technology in the 1910s. In the 1970s, Dr. Robert M. Salter of the RAND Corporation proposed a VacTrain route down the northeast corridor of the United States, but the estimated $1 trillion price tag killed the project. Tunnel-boring technology has improved dramatically since then, and the project is back on the desks of engineers in China, the U.S., and England. Today the cost of a transatlantic tunnel is thought to be closer to $175 billion, which seems downright affordable in comparison.

TOPLESS SANDALS

The only way to get away from it all, to truly relax, to really commune with nature, is to walk around the Great Outdoors totally barefoot. Of course, that's a terrible idea–the Great Outdoors is full of jagged rocks, pine needles, scorpions, and broken glass. Only a moron would walk around the Great Outdoors without any sort of protection on their feet.

Granted, they make those ultra-snug, second-skin-like running shoes that fit around your feet and even have little holes for each of your toes, but those look incredibly goofy, and your feet get hot in those on a hike or at the beach. Instead, you could go with the Topless Sandals. Essentially foot-shaped slabs of rubber, they are foot-shaped slabs of rubber that laboriously stick to the bottom of your feet when you trek through nature. (The manufacturer guarantees that you can remove the sandals when you need to, and that the sticky icky on the shoes will last for up to a year.) So the bottoms of your feet are protected and bound, but the tops of your feet are as free as a bird.

PIGEON-GUIDED MISSILES

Sometimes a dubious notion can come with an impressive pedigree. In 1944 the U.S. National Defense Research Committee, looking to step up the Navy's attack capability against German battleships, engaged renowned researcher and boxer of children B. F. Skinner to help develop a missile-guidance system. Skinner was an unlikely choice for the job, because, while he was undoubtedly a brilliant fellow, he was–literally–no rocket scientist. Rather, his specialty was behavioral psychology.

Skinner had devised a method whereby a missile's flight could be directed by trained pigeons riding in the nosecone. The pigeons watched the target on monitors and were conditioned with food rewards to keep the target centered onscreen by adjusting the missile thrusters with beak-activated switches.

Project Pigeon was scrapped when the Navy decided that its existing mechanical guidance systems were accurate enough for the task. Skinner later complained, "Our problem was that no one would take us seriously"–that is, the idea was rejected simply because it was unconventional. But there were practical considerations as well. Training and sustaining the pigeons was expensive and time-consuming, and the birds were, sadly, not reusable. The project was briefly revived in the early days of the Cold War but, like many a birdbrained scheme, never caught on.

IN-CAR RECORD PLAYER

In 1955 Columbia Records came up with a novel way to sell more records: Install record players in cars. Engineers solved the obvious problem of how to keep the needle on the record while the car rides along with a spring-loaded tonearm. The disc was also twice as thick and heavy as a regular record, which helped keep it from bouncing off the turntable.

Columbia talked Chrysler into making the Highway Hi-Fi an option on all new 1956 models, and produced 21 "Highway" records to go with it (mostly classical and Broadway cast albums). The main problem was that they didn't know their audience: Teenage drivers were the ones who'd want to play records in the car, and they bought rock 'n' roll music. But they didn't buy very many new cars, and the adults who did weren't much interested in a record player—Chrysler abandoned the concept after just two years.

ALUMINUM FACIAL SPA

The Indian sage Swatmarama wrote, "When the breath wanders, the mind also is unsteady." While his words may have referred to breathing techniques during yoga, a more recent implication is that those who invest in the Aluminum Facial Spa will have the steadiest minds around. Produced by the Japanese company Akaishi, this aluminum mask covers its wearer's entire face and uses their own lukewarm breath as a steam facial treatment. The mask fastens around the back of the neck via a Velcro strap using the time-honored hook-and-loop method, and snuggles the jaw–fans of Marvel Comics will find that it falls somewhere between the visages of Iron Man and Dr. Doom. Oh, right, and it's also bright pink.

Although everyone breathes and sweats differently, the recommended usage time of the mask hovers at around 15 minutes per session, with the Amazon.com description for the product–which appears to have been translated from Japanese to English somewhat haphazardly–warning potential buyers, "Please discontinue use if symptoms are itching, and rash appeared in use or after use." Lastly, while most would reasonably theorize that the mask should be rinsed with water after each use, the company also advises not to clean the mask with a washing machine, dryer, hair dryer, or iron.

THUNDERSHIRT

Phil Blizzard had a 10-year-old dog named Dosi who was very friendly and healthy, but terrified by loud noises such as thunderstorms and fireworks displays. Anxiety medicines prescribed by a veterinarian weren't effective, because they took too long to work; Dosi could sense a storm was coming and would already be freaked out by the time Blizzard could give her a pill.

A trainer suggested noise desensitization, so Blizzard played recordings of thunderstorms. But Dosi wasn't an idiot. It didn't work. So what did Blizzard do? He gave Dosi a hug. A permanent hug.

Blizzard developed the Thundershirt—an anxiety-relieving sweater vest for dogs. Simple and easy to slip on, it seemed to cure Dosi of her freakouts immediately. Amazingly, they work for most dogs, and they're widely available in pet-supply stores now. The concept is similar to swaddling a baby and holding them close to calm them down. The Thundershirt does the same thing for dogs (and cats): hugging them when you can't get down there and hug them yourself.

EXTREMELY INSTANT NOODLES

A boiling pot filled with noodles can take anywhere from five to 10 minutes to cook. That's five to 10 minutes too long, depending on how hungry you are. But thank the Flying Spaghetti Monster: Science has finally given us a faster noodle.

In 2013 Royal Chef, a Japanese food company, released Eight Second Spaghetti. Unfortunately, that name is a bit of a misnomer. Pasta cravers still need to wait for a pot of water to boil first. However, once the water's bubbling, the pasta does, indeed, cook in just eight seconds. That's fast enough to make a bowl of instant noodles seem downright slow by comparison. The noodles come in three different varieties: thin, normal, and hefty (they're normal noodles, but there's more of them). Prior to cooking, they look like unappetizing bricks, much like the stuff found in a typical package of ramen.

Eight Second Spaghetti is also spendier than your average container of Cup Noodles. The packets cost between 550 and 580 yen. That works out to over $6 per plate. And another drawback? You still need to heat up some spaghetti sauce, too.

MIRACLE WEIGHT-LOSS PRODUCTS

Spray-on weight loss: CLAmor contains a chemical called Clarinol that's thought to shrink fat cells. When the clarinol-sprayed food is eaten, it reduces fat on the food and fat that's already inside the body. It comes in four flavors: butter, olive oil, garlic, and plain. So what is Clarinol? CLAmor's maker says that it's a naturally occurring bacterium found in the stomach of cows. It's harvested from fried ground beef, and also it doesn't work.

Fat converted to water: A pill called Phena-Frene/MD sold in the mid-1990s claimed to turn fat into water, which was then flushed from the body by peeing it out. One problem: It's physically impossible to turn fat into water. The product bombed, despite citing studies from the California Medical School and the U.S. Obesity Research Center, neither of which exists. Phena-Frene was banned in 1997, and also it didn't work.

Ear-clip your way to a new you: According to Ninzu, the manufacturer of a device called the B-Trim, weight loss could be attained by clamping this Bluetooth-like object onto the ear. Here's how it "worked": The clip put pressure on a nerve ending, which supposedly stopped stomach muscles from moving, signaling to the brain that the stomach was full. This controlled the appetite and resulted in weight loss. The Federal Trade Commission made Ninzu stop selling the B-Trim in 1995 because it didn't work.

SPRAY-ON SKIN

The next time someone (probably your grandfather) starts bellyaching about how "if we can send a man to the moon, how come we can't do [whatever thing he saw on *Star Trek*]?" you can shut him up with three simple words: Spray-On Skin.

Spray-On Skin, developed by Avita Medical Limited, is used in a variety of medical procedures, primarily wound treatment, scar remodeling, and cosmetic surgery, but perhaps most importantly it aids in the treatment and healing of burn victims. And before you even ask, this does not involve keeping spray bottles full of skin at the ready. That would be gross. Cool, but gross.

The process involves harvesting a quantity of healthy skin roughly the size and thickness of one or two postage stamps, which is much less destructive and painful to the patient than a traditional skin graft. The harvested skin is liquified, then mixed with a special enzyme and sprayed onto the areas to be treated. Within a week, an area of skin the size of a stamp can grow to the size of a sheet of paper, although we aren't sure if that's letter, legal, or A4 size.

THE THEREMIN

The sound is familiar, even if the name isn't. The warbling melody at the fade-out of the Beach Boys' "Good Vibrations," the electronic keening that underscores 1950s sci-fi movies–it's a tone somewhere between a slide whistle and a singing saw: That's the sound of the theremin, one of the strangest musical instruments ever created.

Purely electronic, the theremin is unique in that the player never touches it. Two antenna-like capacitors–one controlling pitch, the other volume–protrude from a box that houses radio frequency oscillators. The resulting signal is fed out through a speaker. The device is played with delicate, precise motions of the hands in the air around the antennae. The effect is eerie, as if the player is conjuring music from the ether; in fact, inventor Léon Theremin's original name for his instrument was the etherphone.

A physicist and amateur cellist, Theremin invented his instrument in 1920, more or less by accident, while working for the Soviet government on a device to detect objects through the air (sort of like radar). The Soviets sent him on a European tour to demonstrate the device (and Soviet ingenuity), and he played to packed concert halls across the continent. In 1928 he defected to the U.S., where he stayed for 10 years, setting up a lab in New York City During his decade in New York, Theremin–along with his protegée, Clara

Rockmore—worked to popularize electronic music. Serious contemporary composers like Percy Grainger, Miklós Rózsa, and Dmitri Shostakovich were soon writing works integrating the theremin into the concert orchestra.

Then, in 1938, Theremin was kidnapped by the KGB and put to work in a secret government lab in Siberia, where he remained until 1966. Upon his release, he turned to teaching, living in obscurity. Many believed he was dead. After the collapse of the Soviet Union, Theremin began to travel again, eventually returning to New York, where he was reunited with old friends, including Rockmore.

It's easy to make *noise* on a theremin, but hellishly difficult to make *music*. It requires tremendous mental discipline—concentration, sense of pitch, and muscle memory—as well as daunting physical skill. During Theremin's half-century absence, it was nigh impossible to find instructors qualified to teach the instrument, and the theremin fell into disrepute. Its use by self-taught rock and avant-garde musicians—many of whom employed it primarily as a sound effect—gave it a reputation as a novelty instrument.

But since Theremin's death in 1993, there's been a resurgence of serious interest in the instrument. A new generation of players and composers—including Theremin's grand-niece Lydia Kavina, a virtuoso trained by the old man himself—are writing and performing new music for the theremin, taking advantage of its ethereal qualities.

MORE KITCHEN GADGETS

Egg Cracker. Everybody knows that eggs have extremely fragile shells—that's why we envision Humpty Dumpty as an egg and use the expression "walking on eggshells." They're remarkably easy to crack them on nearly any hard surface. That's why it's difficult to imagine who would ever be in the market for the $10 EZ Cracker. Resembling a garlic press, it has an egg-shaped chamber where you put the egg. Then you squeeze the handle...and it cracks the egg into a bowl, retaining the shell.

Butter Cutter. Cutting softened butter is easy; slightly harder is cutting off a slice of butter just out of the fridge. Not a problem for the One-Click Butter Cutter. Just insert a stick of butter in this $12 plastic doo-hickey, and then squeeze it, and out drops a perfectly square pat of butter.

Hot Dog Hamburger Mold. It's the barbecue attend-ee's dilemma: hot dog or hamburger? Now you don't have to choose, because the Hot Dog Hamburger Mold ($15) makes hamburgers...in the shape of a hot dog. Fill up the hot-dog-shaped mold with ground beef, and you've got a hot-dog-shaped hamburger patty that you can serve on a hot-dog bun. (This product is great for people who bought the wrong buns at the store.)

THE 10-IN-1 GARDEN TOOL

Storage is a common frustration for the weekend gardener: What to do with the hoe, the shovel, the pruning shears, the rake, the scythe, the machete, the grass-maintaining goat, and all the other things that might only come out of the shed once or twice a year. A Japanese inventor, apparently tired of running over his rake every time he pulled into the garage, decided it might be a good idea to combine all his garden tools into a single unit, with space for each tool to be folded away when not needed.

If this sounds like a Swiss Army knife, that's what it looks like, too. Ten tools in one device! A multi-function apparatus for the less-than-constant gardener! The only trouble is that, whereas a corkscrew or can opener is compact enough to tuck into a pocket knife, a shovel or hoe isn't. And a five-foot-long, 20-pound Swiss Army knife handles like a five-foot-long, 20-pound Swiss Army knife. Not exactly the tool for those precision jobs that require a firm grip, a steady stance, or the ability to see the dirt into which you're digging.

BAGELHEADS

Tattoos and piercings are so yesterday. This strange form of body modification is known in Japan, where it's been popular for nearly a decade, as *seerin durippu*, or "saline drip." Technicians insert a needle in the client's forehead and inject about 400 cc of a harmless saline solution, which creates a large bulge just under the skin. The technician then uses their thumb to create an indentation in the middle of the bump. The whole procedure takes around two hours, and the end result looks like a subcutaneous bagel.

"Bagel bumps" are temporary. It typically takes between 16 and 24 hours for the body to absorb all the saline and the forehead to return to its normal shape. Unlike a tattoo or a facial piercing, a bagelhead will turn heads on Saturday night, but fade away by Monday morning.

Japanese journalist Ryoichi "Keroppy" Maeda popularized bagelheads. He's been covering Japanese body modification fads since the early 1990s. At a body mod convention in 1999, Maeda ran into an artist named Jerome Abramovitch, who was experimenting with saline infusions. With Abramovich's permission, Maeda began organizing bagelhead booths for parties and events around Tokyo. The fad took off from there, and more daring fans of the procedure have begun injecting saline into *other* body parts.

GOOGLE GLASSES

Google is probably the world's leading research-and-development company, particularly with its well-funded (but highly secretive) Google X division trying to make sci-fi a reality. Founded by CEO Larry Page in 2005, one of the division's primary directives is to eventually create a direct line between the human brain and consumer technology powered by Google's many online services.

Google envisions a future in which you don't have to type or even look anything up on a computer–just think of what you need or want and it will pop up in front of you. That's the idea behind the company's most promising gadget yet: Google Glasses. They're a pair of transparent glasses hardwired with smartphone technology. As the wearer walks around, the glasses display things in front of them that only they can see, such as directions superimposed on the street, weather reports, or incoming text messages.

The company has sent employees out to major events (such as the 2013 Academy Awards) to demonstrate and test the Glasses in the real world. But the ultimate goal, according to project leader Babak Parviz is to put the system on a pair of contact lenses, which would utilize radios no wider than a few hairs. Then, Parviz theorizes, you could watch videos or even monitor your health with tiny biosensors. This gives new meaning to the phrase "Googling yourself."

THE SOVIET CALENDAR

Rather than follow the decadent, lazy, and inefficient Gregorian calendar used by the U.S. and literally every other country on Earth, in 1929 the U.S.S.R. introduced its own internal calendar. Instead of 52 weeks of seven days each, the Soviet calendar consisted of 72 five-day weeks. That makes 360 days, so the other five days became holidays to mark important dates in Communist Party history.

The purpose of the calendar was the same as the purpose of every other Soviet innovation: to squeeze more work out of the workforce. Instead of having two days off out of every seven days, workers got one day off every five days. On physical calendars, the five days of the week were printed in different colors, and each worker was assigned a color to indicate which day was their day of rest. The new calendar made it easier for factories to remain in operation every day of the week. It also furthered the ideological goal of de-emphasizing Sunday as a religiously ordained day of rest, because religion was frowned upon in the U.S.S.R.

The new system didn't increase productivity much. Machinery in constant use tends to break down a lot, as do people. In 1931 the five-day week was scrapped in favor of a six-day week with a common rest day for everyone. That lasted only until 1940, when the Soviets went back to the seven-day week with a two-day weekend.

THE TRANQUILITY CALENDAR

Many would argue that the single most impressive moment in the history of Earth is when we shot three guys off of it and landed them on the moon. Such an important event marks a new epoch in human history, and as such, we should adjust our calendars accordingly. This is he theory behind the Tranquility Calendar, a system proposed by Jeff Siggins in the July 1989 issue of the science magazine *Omni.*

Here's how it would've worked: This calendar uses as its starting point "Moon Landing Day," or more precisely, the moment (20 hours, 18 minutes, 1.2 seconds) when *Apollo 11* set down on the moon and Neil Armstrong said over the radio to NASA's Mission Control, "Houston, Tranquility Base Here. The Eagle has landed." Dates before and after Moon Landing Day are designated as BT ("Before Tranquility") and AT ("After Tranquility"). The Tranquility calendar is a perpetual calendar with 13 months of 28 days each. That adds up to 364 days. The anniversary of Moon Landing Day, called "Armstrong Day," serves as the 365th day. It is a "blank day" that is not part of any week or month. In leap years, a second blank day, called "Aldrin Day" after Neal Armstrong's crewmate Buzz Aldrin, is also added.

But old habits die hard. People are insanely loyal for decades to their brand of beer—good luck getting them to change how they view time.

THINGS INVENTED BY THE PROFESSOR ON *GILLIGAN'S ISLAND*

- Lie detector (made from the ship's horn, the radio's batteries, and bamboo)
- Bamboo telescope
- Jet-pack fuel
- Paralyzing strychnine serum
- "Spider juice" (to kill a giant spider)
- Nitroglycerine
- Shark repellent
- Helium balloon (rubber raincoats sewn together and sealed with tree sap)
- Coconut-shell battery recharger
- Xylophone
- Soap (made from plant fats, it's not really so far-fetched)
- Roulette wheel
- Geiger counter
- Pedal-powered bamboo sewing machine
- Pedal-powered washing machine
- Keptibora-berry extract (to cure Gilligan's double vision)
- Pedal-powered water pump
- Pedal-powered telegraph
- Hair tonic
- Pedal-powered generator
- Various poisons and antidotes
- Pool table (for Mr. Howell)
- Lead radiation suits and lead-based makeup (for protection against a meteor's cosmic rays)

ICE-BASED AIRCRAFT CARRIER

In 1943 Geoffrey Pyke, a science advisor to the British military, made a radical proposal: Build unsinkable aircraft carriers out of ice. His reasoning: Glaciers are made of ice, and they are virtually unsinkable. Building hulking aircraft carriers out of ice would also be cheaper than using metal and other materials, and it would protect Atlantic convoys against attacks from German U-boats. "Project Habbakuk" was on. (It was named for a biblical prophet who promised unbelievable things to his adherents.)

The scale of these floating landing strips would be immense: 4,000 feet long (more than three-quarters of a mile), with 50-foot-thick hulls and displacement of two million tons of water. And since they were to be made out of ice, the vessels would have been virtually unsinkable, yet easy to repair if damaged by torpedoes.

A 1,000-ton prototype was being built on Patricia Lake in Alberta, Canada, but the project was abandoned when the British were informed that completing it would cost $70 million (about $1.2 billion in today's money) and take 8,000 people working for eight months. The refrigeration units were turned off and the hull sank to the bottom of the lake, where it eventually melted.

HAMSTER SHREDDER

Hamsters aren't the best pets in the world. They're glorified goldfish, but much messier. Plus, sunflower seeds and sawdust aren't getting any cheaper these days. If only there were a way to make these furballs earn their keep. Now there is…sort of. In 2007 London-based consultant Tom Ballhatchet created the Hamster Shredder while working toward his master's degree in industrial design. The ingenious device will help your pet burn calories while helping you destroy documents that might contain personal information ripe for picking by identity thieves.

The Hamster Shredder is a cage that contains a modified hamster wheel connected to a paper shredder. On average, a hamster has to run for 45 minutes in order to eliminate one standard 8.5 x 11 sheet. The paper shreds can then be used for hamster bedding.

Ballhatchet told a reporter for The *Telegraph* that he wanted to "capture people's imagination while addressing issues of topical concern such as climate change, recycling, and identity fraud."

SHOTGUN WITH SHOT GLASS

Drinking isn't the healthiest thing for you, but it's definitely healthier than shooting yourself in the mouth; one may eventually kill you, the other will do so promptly. And sure, both drinking and shooting are recreational activities enjoyed by millions each, but that doesn't mean anybody wants to combine the two.

The "Simulated Firearm With Pivotally-Mounted Whiskey Glass," as its 1969 patent officially names it, seems to derive its existence entirely from the fact that the word "shot" has both liquor and gun connotations.

The invention: It's a novelty gun with a shot glass mounted on top of it, on a moving pin. When you pull the trigger–BANG! The force of the gun propels the shot of whiskey–or whatever you've filled the shot glass with–into your mouth. Or your nose. Or your eyes. Liquor *burns*.

PEDAL-LESS RUNNING BIKE

Combining the cardiovascular workout of a brisk jog, the maneuverability of a bicycle, and the pointlessness of a kick scooter, the Fliz Bike—from the German *flitzen*, "to whiz or dash"—is a new innovation in human-powered locomotion, with a very old design concept. It has two wheels, brakes, and handlebars, but no pedals—and no drive chain. Like the first two-wheelers of the 1820s, it depends entirely on footpower, gravity, and inertia.

Made of lightweight glass-and-carbon-fiber laminate, the bike is a wheeled step-in frame. The rider's upper body is leaned forward at all times, secured by a unique five-point harness system that is custom-crafted for each user. You don't ride the Fliz—you wear it.

Riders accelerate the Fliz by running, and when they reach a good coasting speed, lift their feet from the ground and into the footrests on the rear frame. The suspended harness and prone posture let users "fly" above the pavement, Superman-style. A Fliz may never win the Tour de France, but for whizzing around town, it's a dashing blast.

CUDDLEBOTS

From security blankets to virtual pets, there's no limit to our penchant for developing attachments to lifeless things we secretly kind of imagine are really alive. And because scientists love nothing more than pinpointing our darkest foibles and exploiting the dickens out of them, we now have Cuddlebots.

"What are Cuddlebots?" you ask, because you are human, and you would very much like to own something that sounds like it will give you affection without needing you to feed it or invest in its eventual half-hearted attempt at earning a liberal arts degree. And although at the moment, in its prototype stage, it kind of looks like a furry loaf of bread, that's basically what Cuddlebots are designed to do. The creator says that "the ultimate goal of this work is the design of more emotional, potentially therapeutic machines that can help people feel better." In layman's terms, the Cuddlebot is capable of understanding different types of touch—as well as learning who's doing the touching—thus achieving "social human-robot interaction through affective touch," nerdspeak for "you will grow to love your cuddly gadgets more than your family."

SECURITY GERBIL

In the 1970s, the super-secretive MI5 wing of British National Security (that nation's equivalent of the CIA) came up with what they thought was a secretive, barely noticeable system to detect suspected spies and terrorists as they tried to pass through the country's airports: trained gerbils.

It's not *that* crazy of an idea. Gerbils are able to detect increased adrenaline levels in human sweat when people are nervous. The MI5 wanted to set up security queues near a row of fans that would blow air into a hidden gerbil cage. The gerbils were trained so that if they smelled that extra bit of adrenaline wafting in, they would press a lever that would alert security. What's in it for them? The lever also gave the gerbil a treat for a job well done.

So how many terrorists did the gerbils catch? Not even one: The program was abandoned after the researchers discovered that the gerbils failed to notice any difference between people who were nervous because they were spies, or merely passengers who were afraid of flying.

THE INTERROBANG

Unlike most everything else in this book, the interrobang isn't a physical object—but it is weird, it is an invention, and it attempted to solve a problem nobody really knew they had until an invention came along to solve it.

Punctuation has been pretty much fixed in our lifetimes—we all know (or attempt to use correctly) all the various commas, semicolons, apostrophes, and whatnot. All have specific purposes. But what about when you're writing a story or a letter and you need to angrily ask a question? You do something like this: "What is the meaning of this?!" It works—yes, it uses two different punctuation marks, but the meaning is conveyed.

American Type Founders thought that using two different marks looked clunky. So, in 1967, they attempted to introduce the interrobang into common English usage. It's a question mark with an exclamation mark laid over it.

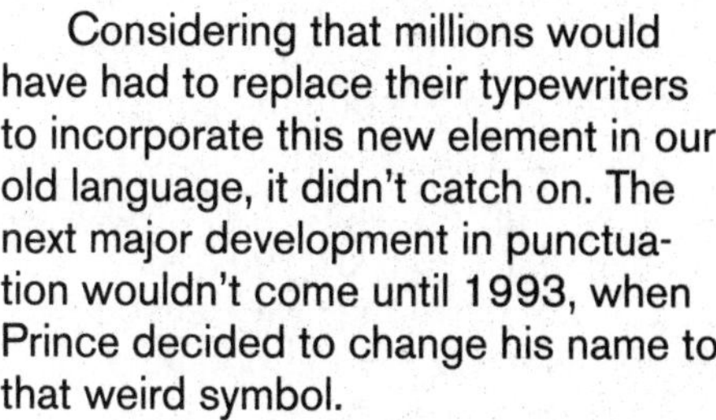

Considering that millions would have had to replace their typewriters to incorporate this new element in our old language, it didn't catch on. The next major development in punctuation wouldn't come until 1993, when Prince decided to change his name to that weird symbol.

THE NUTRITION PATCH

There are all kinds of ways to get vital nutrients into your body: a feeding tube, injecting them into your veins, or if you're old-fashioned, eating food. In a probably misguided attempt to improve on these methods, researchers for the U.S. Army are developing a quick and easy nutrition patch, referred to as the Transdermal Nutrient Delivery System. The idea is that the patch would stick to the skin like a bandage and provide a steady dose of vitamins throughout the day, so the wearer could remain biologically alive and nutritionally fortified (but probably extraordinarily grumpy and hungry) without eating.

Though the nutrition patch is intended for military use, army researchers suggest it could also be used by miners, astronauts, and other workers whose jobs require too much extended effort and concentration to allow a proper lunch break. (Others whose jobs meet this description: brain surgeons, stay-at-home moms, and strippers).

You can already buy a low-rent version of the patch on the type of discount health product websites where you can also get off-brand colon cleanses, pills that claim to regrow hair, and other products you've read about in this book. Or, you could join the army and wait until 2025, when the real Transdermal Nutrient Delivery System is expected to become available. Or you could just, like, eat a sandwich.

SPEAKER VEST

There are so many variations on this design, from so many different entrepreneurs, that it's probably unfair to count the speaker vest as just one invention. But all these wearable personal audio systems aim to solve the same social problem—namely, how best to annoy every living soul within 50 yards, now that those comically oversize boomboxes are out of fashion.

The speaker vest was conceived to serve the practical purpose of allowing motorcyclists to enjoy music while they ride, without using headphones (which dampen ambient noise, making wearers less aware of possible danger around them; for this reason, it's illegal in most places to operate a vehicle while using them). But what began in the 1980s as a low-power, two-speaker rig tethered to the bike's dashboard has become something else entirely.

Stereo enthusiasts are prone to a more-is-more mentality, adding more speakers, more watts, and more output until they're left with something that could actually do them harm. One iteration of the speaker vest, custom-built and sold online, has a half-dozen cone speakers built into the torso and an eight-inch subwoofer—powerful enough to shake the user's spine loose from his brainpan—mounted on the back. Forget traffic hazards; this thing is dangerous in itself.

GREEK FIRE

In the 3rd century A.D., the Roman Empire had become too big to manage, and so it split into a western half ruled from Rome and an eastern half ruled from the Greek-speaking city of Constantinople. While the western empire fell in the 5th century, the eastern part soldiered on as the Byzantine Empire and lasted another 1,000 years. Over time, the Byzantines lost ground to the Arabs, whose territory expanded in the 7th century until much of the eastern Mediterranean fell under their sway. Constantinople itself came under siege from the Arabs between the years 674 and 678, and might have fallen had it not been for the Byzantines' secret weapon: "Greek fire."

Greek fire was most likely invented by a Syrian engineer named Kallinikos, although it was probably also the result of generations of experimentation. What is definitively known is that it gave the Byzantines a huge advantage in naval battles. Unfortunately, the process by which it was made was kept a strict secret, and the recipe has been lost to history. (We have only written accounts from witnesses.) It was a sort of burning oil that was ejected by siphons onto enemy ships, accompanied by smoke and a sound like thunder. The devastating thing about Greek fire was that it couldn't be extinguished with water, and any ship doused with it would continue burning until it was a floating pile of cinders.

We do know it was a liquid, that it burned in water, and, by some accounts, it was ignited by water. It could be put out only with sand, strong vinegar, or old urine.

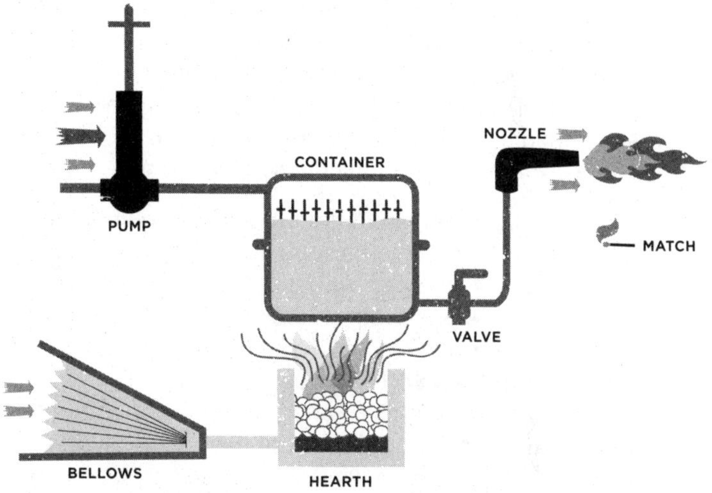

One theory is that Greek fire contained calcium phosphide, which ignites in contact with water. However, experiments have shown that it doesn't burn long enough, or with enough intensity to do the damage described. Another idea: Greek fire could have contained saltpeter, an ingredient in gunpowder. But saltpeter was unknown in Europe before the 13th century. Historians now more or less agree that Greek fire was probably some kind of petroleum made from crude oil, possibly thickened with resin to increase its intensity. If so, it was similar to the napalm used to devastating effect in the Vietnam War.

CHARLIE SHEEN'S CHAPSTICK DISPENSER

Celebrities have always been America's most precious resource, and our chiefest source of good ideas. One of Hollywood's greatest innovators is award-winning actor and notable newsmaker Charlie Sheen, who was awarded a U.S. patent in 2001 for a new way of dispensing lip balm.

The main problem with the tube form known to all who have suffered from chapped lips, Sheen believed, is that the pesky little cap would keep falling out of the user's hands. He and partner Rodger D. Thomason devised a new tube with a slidable lid that would remain on the dispenser. The lid served a dual purpose. When opened, it turned into a finger cradle, so the balm could be applied more easily while wearing gloves, perfect for use in winter by skiers and professional snowball fighters.

But that's not the full extent of the invention's improvements. The dispenser did away with the rotating wheel at the bottom, replacing it with a slider on the side. The new tube also had indicators on the side, so users could see how much lip balm was left without having to open the (frequently lost) cap and peek inside.

As if they were all producers on a comedy show Sheen left in a grand and destructive fashion, no major lip-balm makers have yet adopted the new technology.

ANTI-GRAVITY BOOTS

Michael Jackson may have been the greatest entertainer of his generation, but for many, his adult life seemed like one big invention, from his cosmetic transformations to home amusement park to...

Still, his artistry was sacrosanct, with many dazzling moments both musical and visual, particularly the Moonwalk and that thing in his 1988 music video for "Smooth Criminal," when he and his dancers repeatedly leaned forward, far beyond the human body's normal center of gravity. How did they do that? The choreography was assisted by an elaborate "hitch" mechanism that enabled the dancers to attach the soles of their shoes to a section of the stage. When they moved their feet forward in the shoes, inserts in their heels—as well as the floor beneath them—tilted forward at an angle up to 45 degrees, allowing the dancers to achieve that seemingly gravity-defying lean.

Jackson and some of his associates actually patented the process, necessary for when he took his "Smooth Criminal" moves on tour, so as to build the mechanisms into his traveling stage.

THE FIRST FLYING CAR

The grand triumvirate of futuristic stuff of Tomorrowland is the video phone, food in pill form, and flying cars. Well, we've got Skype and nutrition patches (see page 212) to account for the first two, so why don't we have flying cars yet? Because we already did...in 1946.

In the 1930s and '40s, carmakers Daimler, Benz, and Ford each spent millions of dollars and countless man-hours researching the concept of a flying car, but ultimately all came to the same conclusion: A vehicle that could both drive on the ground and fly would do each so poorly that no one would buy one, nor believe they were safe. Then, in 1945, an aircraft engineer named Theodore Hall designed a regular car so small and light that you could attach an airplane engine, wing, and tail unit to the roof...and he made it fly.

Hall planned to make the airplane units available at every airport in the country, where driver-pilots could rent them for a small fee. He sold his idea to the airplane manufacturer Convair, which marketed it in 1946 under the name ConvAirCar. This might well have changed the face of the automobile industry forever. Unfortunately, in 1947 a Convair pilot miscalculated the amount of fuel he needed and was forced to crash-land his ConvAirCar on a dirt road, shearing off the plane's wings in some trees. The adverse publicity doomed the project.

STUD SPECTACLES

It seems like everything that begins as edgy eventually ages and finds a more sober and boring purpose in life. Bands like The Who and Led Zeppelin helped define an era of rock 'n' roll rebellion and youthful excess, yet their music was eventually used to sell four-door luxury sedans. More recently, those who got piercings in the '90s have become accountants, dental hygienists, and marketing executives. For those who still have rings, studs, and gauges in places besides their ears, the Stud Spectacles can put them to use.

Stud Spectacles (stud as in the jewelry term, not as in your reputation, chief) offer a practical use for those eyebrow piercings your cousin Brittany got that summer when she went to Lollapalooza. Instead of traditional eyeglasses that rest on one's ears and nose, this specially-crafted frame hangs from eyebrow piercing studs. If you're looking for a pair of glasses for cousin Randy with the ear-hole gauges, you're out of luck, but there is another version of this invention that hangs from a single piercing through one's nose bridge.

Despite the name, it isn't clear that these glasses will enhance anyone's studliness or sex appeal. That is, unless, you find "practical eyewear that society at large isn't yet ready for" to be studly.

NEUTICLES

Even though they may nott understand many English words beyond "outside," "food," and "baaaaaaad," we use the word "neuter" or "fix" around our dogs as a euphemism for what the medical procedure really is–castration–so as to prevent them from making puppies with the flirty poodle next door.

Apparently, inventor Gregg Miller decided that dogs, the creatures who happily eat garbage and even more happily lick themselves where they've just been neutered or fixed, were self-conscious and apprehensive about lacking the physical, external machinery post-surgery.

So Miller created Neuticles, a portmanteau of the words "neuter" and "testicles." Neuticles are plastic, decorative testicles you can have surgically implanted in your dog where his real testicles used to be. This way, it stands to reason, he won't feel bad and will look like all the dogs in the neighborhood who didn't get their genitals lopped off.

Neuticles are available in a number of sizes for dogs (and cats!), resemble a tiny breast implant, and cost a little over $100 a pair.

BED STRAIGHTENER

You know how when you make your bed, you do that thing where you kind of toss the bedspread or comforter up in the air so that it sort of inflates like a balloon and then spreads out as it falls, and then you just kind of straighten it out? The whole process takes about 10 seconds...but what if there were an unnecessarily complicated and bulky contraption you could put under your bed that could do that one, exact thing for you?

There is! It's the Bedding Straightening Mechanism, patented in 2008 by a team of inventors from Chile. It consists of an air compressor (or, alternately, a turbine) placed inside of a soundproof box under the bed. When turned it on, it shoots air into the bedspread (and the rest of the bedding, for that matter), which makes it do that thing where it goes up in the air, fluffs up, and falls back down on the bed slightly straighter than before. It's still up to you to finish making the bed all the way, though.

SPRAY-TAN TENT

So you love the spray-tan look, but you hate going to a salon where some skeevy dude making minimum wage ogles you in your swimsuit while he's hosing you down with bronzer? Maybe you'd rather spray-tan at home, if you don't mind investing in your own airbrush. But after scrubbing tanning-solution residue off the walls of your shower a few times, you'd realize that the real advantage of using a salon is that somebody else cleans up afterward.

But fear not! The Spray-Tan Tent is a lightweight nylon contraption that pops up into a cubicle four feet wide and seven feet high. Available in a range of colors and equipped with a clear vinyl roof, this portable sanctuary maintains your modesty while keeping the mess of sunless tanning contained. Best of all, it folds down to the size of a large handbag when not in use.

Of course, the tent alone—not even factoring in the cost of the airbrush, pigment, and ventilation fan—could set you back as much as $100. Because this is totally a legitimate, professional-grade accessory, and in no way simply a child's pop-up play tent repurposed and marked up to a ridiculous margin for a new market.

WIND-POWERED MOLE CHASER

Tired of moles and gophers ruining your lawn and garden? Instead of using poisons and traps to kill them, you can annoy them off your property with the $23 Mole Chaser. It looks like a weathervane or windmill mounted on a long metal pipe that you hammer into the ground. The spinning blades vibrate in the wind, sending pulses down the pole and into the soil–which drive the underground varmints nuts, forcing them to move elsewhere. The idea may sound farfetched, but the devices have been around for years, and gardeners swear by them.

Each Mole Chaser is effective across 10,000 square feet of ground. And if there isn't a lot of wind in your area, you can use battery- and solar-powered versions, which cost $25 to $40.

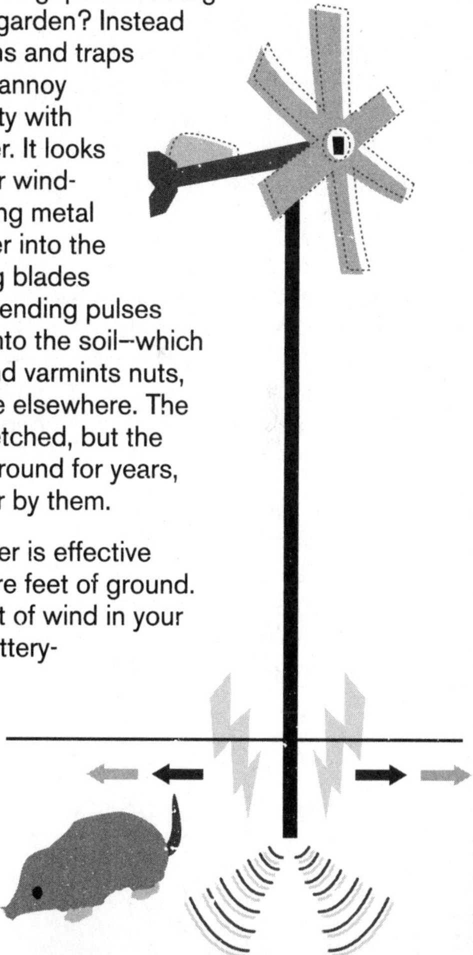

SKITTLES SORTING MACHINE

One of the things that makes obsessive-compulsive disorder such a drag is that it takes so much time to get anything done, especially the arduous task of sorting through bags of multicolored candy to separate the pieces by individual color. Thanks to an enterprising fellow by the name of Brian Egenriether, we now have the Skittles Sorting Machine.

The machine, which looks something like an old-fashioned food processor, uses BASIC Stamp 2 and 3 servos for actuation. An IR LED and phototransistor are used to stop the turnstile in position, which allows the color-sorted candies to drop into different bowls. If you understood any of that, congratulations, you deserve some Skittles!

Just don't expect to eat them right away. A video Egenriether posted online of his machine in action reveals that it sorts Skittles at a rate of 37 per minute, or about as quickly as you could do it by hand. Still, it does free you up to perform more important tasks, like hand-sorting all your M&Ms.

THE NEARLY IMMORTAL SANDWICH

For decades, American soldiers have made do with MREs (short for "meal, ready-to-eat"). While these kits are necessarily easy to prepare and consume in the field, they're notoriously unappetizing and typically contain stuff like bland crackers and freeze-dried meatloaf. However, they've been steadily getting tastier in recent years because military culinary scientists (a thing) have been researching new ways to keep foods fresh for longer and longer periods of time. One of their latest developments is a sandwich with a two-year shelf life.

Bacteria, mold, and moisture can turn a yummy sandwich into an icky mass of glop in just a few days. After studying stuff like honey and salt, which are exceptionally good at retaining moisture, the scientists got to work on a sandwich that does the same thing. In 2011 they unveiled a cutting-edge hoagie that can stay fresh for up to 24 months. It locks in moisture, which prevents the bread from going stale. It's also stored in an air-sealed container with a packet that absorbs water molecules in the air. Soldiers say that the new sandwiches are a definite improvement over the ones they're used to finding in their MREs, despite the strong possibility that that sandwich is older than their children.

TURTLE SUBMARINE

We generally think of the submarine as a modern invention, but it actually dates to the American Revolution. In 1775 Yale graduate David Bushnell designed a one-man submarine, which he called the *Turtle*. About six feet tall and three feet wide, and shaped like an egg, the *Turtle* consisted of two wooden shells waterproofed with tar and held together with steel loops, like a barrel. Inside were a variety of controls, including a foot pedal that cranked the propeller and a hand drill for boring into the hulls of enemy ships (damn the torpedos…because they hadn't been invented yet).

The peculiar submersible attracted the attention of Benjamin Franklin, who knew a good invention when he saw one and recommended it to General George Washington. Washington was skeptical, but nevertheless provided funds for the *Turtle*'s completion in 1776. Sergeant Ezra Lee was given the daunting task of pedaling the craft through the waters of New York Harbor. His mission: attach a keg of explosive powder to the hull of the British warship H.M.S. *Eagle*. Unfortunately, because of copper plating, Sgt. Lee wasn't able to bore into the hull as planned, and he was forced to abandon the attempt, and let his makeshift torpedo float away (it later exploded in the East River). A second attempt to blow up another British ship failed, and General Washington decided to abandon the project, although he praised it as a "work of genius."

THE WORLD'S TINIEST CAR

....isn't very much fun to drive. In February 2013, Jeremy Clarkson, who co-hosts the BBC program *Top Gear,* unveiled the P45, which he dubbed "The World's Tiniest Car." Clarkson allegedly designed the hybrid vehicle, taking inspiration from the Peel P50, a three-wheeled British "microcar" that was released in the early '60s. It's still considered the smallest production car of all time. Like the P50, the P45 was built to be both street-legal and as tiny as possible.

Clarkson's creation resembles the offspring of a Jet Ski and a Rascal scooter. The P45's roof consists of a helmet. The closest thing to a windshield? The helmet's visor. There are no side doors, and the contraption makes a Mini Cooper look like a Humvee. But the driver is enclosed, just barely, with their head popping out, so it technically isn't a motorcycle.

Nevertheless, with its 1.7-liter fuel tank and two-stroke 100cc engine, the P45 gets amazing gas mileage. As Clarkson proved during his test drive, it's also a fantastic car for weaving in and out of traffic on the UK's narrow roadways. The driver needn't get out of the P45 to fill up the tank at gas stations either. On the other hand, the P45's not so great with potholes, and driving it on freeways can be downright terrifying. As of press time, exactly none of the car industry's major manufacturers were clamoring for Clarkson's blueprints.

HORSE CAR

Automakers in the 1890s faced major obstacles in winning public acceptance of their newfangled contraptions. People still trusted horse-drawn vehicles over the unreliable early autos, and because automobiles occasionally spooked the horses they passed on the road, many people considered cars a public nuisance. The obvious solution: Combine this newfangled contraption with the old-fashioned reliable appearance of a horse. By which we mean, "Tack a huge fake horse on the front."

In 1897 carmaker Joseph Barsaleux built a car that had a combustion engine and all that, but looked like a carriage, with a full-size replica of a horse in front. The horse camouflaged a fifth wheel that provided power and steering, literally pulling the rest of the vehicle along the road. The driver steered the vehicle using a brace and bit attached to the faux horse's mouth for that familiar horse-controlling sensation. However, by the dawn of the 20th century, the public was getting used to automobiles, and people were replacing their carriages and horses with them. There was no longer a market for Barsaleux and his weird mannequin horse car.

PEEPING THOMAS

There are lots of products out there designed to make it look like someone is home when you're on vacation (or even just at work for the day), so as to thwart burglars who might be casing the joint. Automatic light timers are one example. Another is the Peeping Thomas, invented by Terry Kirby after he walked in on three men trying to burgle his Texas home.

Kirby says he got the idea for Peeping Thomas almost immediately after the robbery. He went over to his grandmother's house to stay for the night; when he knocked on the door, she put an index finger through two slats of mini-blinds and peered out to make sure it was him.

Peeping Thomas is an automated mini-blinds peering device, making it look like somebody is suspiciously leering out the window at you, the potential criminal. How it works: A base holds up a two-foot-tall metal pole. A hook in the middle of the pole goes beneath an eye-level mini-blinds slat. A timer in the device moves the hook, and opens the mini-blinds just a hair, at intervals of 5, 10, 20, or 30 minutes. That gives the impression that someone is home...and constantly checking.

TWO HOMEMADE INSTRUMENTS

Pencilina: Invented by Brooklyn musician Bradford Reed, who used to play the zither in Blue Man Group, the pencilina features two wooden boards that serve as guitar necks, mounted atop legs, like a steel guitar. One board is strung with six guitar strings, and the other has four bass strings. Six electric pick-ups gather and send the sound to an amplifier. Double guitar! But wait, there's more: Wedged under the guitar strings are two drumsticks that can be moved to change string length and, thus, pitch. The pencilina creates an array of almost otherworldly sounds that the *Village Voice* said "sounds like Jimi Hendrix and Buddy Rich playing 'Dueling Banjos.'"

Glockenmundharmonika: This instrument was patented in 1908 by New Jersey inventor Ernst Koch, and its mouthful of a name means "bells mouth harmonica" in German. Unlike a normal harmonica, it didn't have any reeds–it had 22 little bells mounted inside. When you blew into the Glockenmundharmonika, tiny spring-loaded hammers would strike the bells. It's unknown if any were ever made, because none are known to exist today. There no recordings either, so nobody's exactly sure what it sounded like, although you can imagine it sounding "tinkly," if it actually worked.

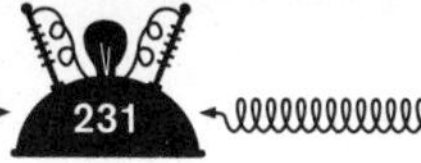

THE ONLINE CHICKEN PETTER

Ever feel the need to pet a chicken, but there just wasn't one handy? Good news: The University of Singapore has invented the Touchy Internet System. "We understand the perceived eccentricity of a system for humans to interact with poultry remotely," says developer Adrian Cheok. "But this has a much wider significance." The device will eventually allow people at zoos to scratch otherwise dangerous animals, such as lions and bears.

Users pet a chicken-shaped doll that's hooked up to their computer, while watching a webcam image of a real chicken on the screen. Sensors on the doll relay the petting location to another computer, which then activates tiny motors in a lightweight jacket that the real chicken wears. The motors' vibrations mimic the sensation of being petted exactly as the user at home is petting the doll. "This is the first human-poultry interaction system ever developed," says Professor Cheok.

AMAZING TOILETS

In 1991 *Saturday Night Live* aired a fake commercial for an imaginary product called "The Love Toilet"—a two-person toilet for people so in love that they never want to be apart, even when they have to use the facilities. Like a Victorian love seat, the side-by-side toilets faced opposite directions, so the lovers could stare into each others' eyes. In a case of life imitating art, the **TwoDaLoo** is now a real item, available for purchase for only $1,400. The only difference between the real TwoDaLoo and the fictional Love Toilet: The TwoDaLoo has a "privacy" bar separating the two commodes (as if that's an issue).

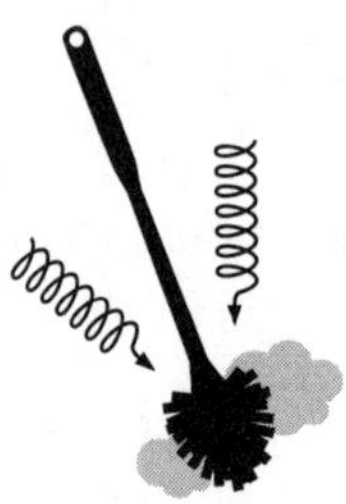

The **Great John** is the first toilet, says the manufacturer, made specifically for "modern Americans." By which they mean it's an extra-large toilet made for extra-large people. Invented by the Great John Toilet Company, the Great John can reportedly accommodate any person up to the weight of 2,000 pounds. The base is wider than a conventional toilet's to provide extra support, and it connects to the bathroom floor with four anchors instead of the standard two. The seat provides 150 percent more "contact area" than a normal toilet (as well as offering side wings to prevent pinching if flesh still hangs over the larger seat).

Lots of people want to teach their cats how to use the toilet (see page 122); it goes the other way with the **Compost Toilet**, essentially a litter box for humans. According to the World Toilet Organization, a legitimate, ultra-serious trade group based in Singapore, this Chinese toilet is a steel box filled with sawdust. It has a microcomputer that senses when the box has solid waste in it, and a mechanical arm that rotates the sawdust, burying the waste, which the company says can later be used as organic fertilizer, which you can then use to grow vegetables to turn into poop later, thus completing the circle of life. The device stays at a constant temperature of about 120°F, hot enough to make liquid waste evaporate, and has specially formulated low-odor sawdust that needs to be changed only once a year, but you can change it more often if you want to.

Shop while you plop! Twyford, a toilet manufacturer in Cheshire, England, created the **Versatile Interactive Pan** (VIP), a toilet that analyzes your urine and stool samples for dietary deficiencies, compiles a shopping list of needed nutritional items, then e-mails your local supermarket to order the foods. "If, for example, a person is short on roughage one day," says Twyford spokesperson Terry Woolliscroft, "an order of beans or lentils will be sent from the VIP to the supermarket and delivered the same day." The toilet can also e-mail a doctor if it detects health problems. Added bonuses: The seat is voice-activated, and the toilet flushes automatically. (That is one pushy toilet.)

GIVE IT A FRY

Enterprising food scientists have recently unleashed amazing new things at state fairs around the U.S., deep-frying things never before thought eligible for deep-frying.

- **Bubblegum.** Marshmallows flavored with bubblegum essence, then battered, fried, and sprinkled with Chiclets.

- **Butter.** Balls of butter are frozen, then breaded and fried.

- **Chicken-Fried Bacon.** The bacon is drenched in buttermilk, then breaded and deep-fried.

- **Fried Pop-Tarts.** It's three breakfast treats in one! It's a Pop-Tart stuffed into donut batter, then fried up like a donut and topped with Fruity Pebbles cereal.

- **Fried Salad.** Salads don't have to be healthy. This is a tortilla filled with spinach, ham, cheese, chicken, tomatoes, and lettuce, then deep-fried and served on a bed of lettuce that no one will ever eat.

- **Fried Frito Pie.** Frito Pie is a southern favorite: a bowl of Fritos corn chips topped with chili, cheese, and onions. The fried version takes balls of chili, cheese, and onions, and rolls them in a breading made with crushed Fritos.

- **Fried Chicken Skin.** No meat necessary.

WINGSUIT

A company called Phoenix-Fly (whose motto is "Human Flight Innovations") sells several versions of what it calls a Wingsuit. The Wingsuit is, essentially, a voluminous full-body tracksuit that fans out in the wind and allows you to hurl yourself out of airplanes and off of cliffs without dying (usually). As the name suggests, most Wingsuit models have large flaps of fabric along the sides that look like wings, eliminating the need for a silly parachute.

The Wingsuit website explains that to even be eligible to purchase a Wingsuit, you have to have completed at least 200 skydives. Ominously, it also indicates that several people have died "so far" in Wingsuit-clad jumping mishaps.

Prices vary; $580 gets you the most basic model of Wingsuit, which looks like a workaday nylon track suit, while for $1,690 you can upgrade to the deluxe Vampire model, which is custom-made for each buyer and looks sort of like what you'd get if you sewed yourself into a camping tent. But hey, if you're the type who doesn't mind jumping off of a skyscraper without a parachute, you probably also don't mind looking like a flying squirrel while doing it.

HEELS ON WHEELS

High heels are uncomfortable, they're awkward, and there's a good chance you'll break your neck while wearing them. They're also not a product of the 20th century–they date to ancient Egypt. Those ancient shoes, which were made out of leather and held together with intricate laces, were all the rage among the higher classes circa 3500 B.C.

But one thing the Egyptians most certainly didn't have back then were training wheels for the clunky shoes. While you probably won't find a pair of "High-Heel Training Wheels" at the nearest Lady Foot Locker, photos of homemade versions have been passed around online for years. Most of the images feature conventional shoes with repurposed toy-car wheels attached to the heels, offering support for those learning how to walk around in the godforsaken things.

Kenji Kawakami, the magazine editor responsible for the Japanese art of *Chindogu* (see page 43), in addition to plenty of ridiculous inventions, may have been the first person to dream up the silly but brilliant devices. While you would think that every young and fashionable gal on the planet would have a pair in her closet, these shoes have yet to become commonplace. Why? Well, among other reasons, the wheels would probably only work well on perfectly smooth surfaces. And stairs would be a challenge.

BEERBRELLA

There's really nothing better than kicking back on a nice day with a tall, frosty beer, either in the backyard, on the beach, or on the porch. The problem with that is that while the sun warms you up nicely, it also warms up your bottle or can of beer, which isn't nice.

It's not like there aren't solutions to this problem. 1) You could drink your beer really fast, but too much beer + sun = naps. 2) The ever-popular foam "beer koozie" wraps around a can and keeps a beer cold, but really only because it blocks the warmth of your hand. It also only works for cans. 3) The Beerbrella.

Patented in 2003, and available in lots of novelty catalogs, the Beerbrella is a pint-size umbrella with a beer-bottle-size plastic clip at the bottom. Simply clip it onto a tall one, and the umbrella shrouds your beer in darkness, away from the warming and beer-ruining rays of the sun.

CLOCKY

Alarm clocks are jarring and annoying, and they brazenly remove us from the deepest sleep. They're a tool we use to wake ourselves up and stay on schedule, but it's extremely easy to turn them off or hit the snooze button if we don't like waking up at the time we've told them to wake us up. The morning battle with the alarm clock is a low-stakes game of man vs. machine, and man will always win.

Unless, that is, the man (or woman) is honest enough with themselves about their snooze-button addiction to buy Clocky. An alarm clock outfitted with two small baby-stroller tires, Clocky rolls off the night-stand and onto the floor at the pre-set alarm time, and it's up to you, in your just-awakened state, to it track down, somewhere in the bedroom and turn it off. And then you're up for good.

Clocky was invented by MIT graduate student named Gauri Nanda, who was inspired to come up with a better alarm clock when she was once two hours late for a class because she repeatedly hit her snooze button. She needed a challenge that would make her spring from her bed and wake her up with some activity. Nauri ultimately designed Clocky for an inventing class.

LADY GUNS

The Chicago Protector Palm Pistol. This gun was disk-shaped, so it looked like a ladies' compact and could be hidden or overlooked in a lady's hand, like a ladies' compact. This seven-shot .32-caliber revolver had a small barrel sticking out of one end and a squeeze trigger at the other end. (The bullets and hammer were inside the disk.) When trouble threatened, it was easy to palm the gun, with the barrel poking out between the index and middle fingers. To fire the weapon, all the user had to do was squeeze her fist. Nearly 13,000 of the guns were made in the 1890s. (Today they're worth $2,000.)

The Frankenau Combined Pocketbook and Revolver. In 1877 Oskar Frankenau of Nuremberg, Germany, received an American patent for a four-shot revolver concealed in a special compartment of a 4" x 2½" metal (but leather-clad so as not to look weird) purse-style pocketbook. All a lady had to do to use it was release a hidden trigger on the bottom of the purse and fire away. According to Frankenau, "The advantage of such a combination for travelers and others will readily be perceived, as it forms a convenient mode of carrying a revolver for protection, especially when attacked, as the revolver may be fired at the robber when handing over the pocket-book."

INSTANT TV COLOR SCREEN

By the mid-1960s, the Big 3 networks—ABC, CBS, and NBC—had nearly completed their gradual transition to broadcasting in color. Not that everybody had a color TV yet. Heck, not everyone even had a TV yet. Millions of people weren't ready to spend another few hundred bucks (in 1960s money) on a color TV when they had only relatively recently bought a black-and-white one, also for hundreds of dollars in 1960s money.

The cheap, bordering-on-low-rent solution: the Instant TV Color Screen (cost: $1). Just place it on a black-and-white TV screen and enjoy instant color. Too good to be true? Of course it was. It was a multicolored sheet of thin, clingy plastic. The sheet was separated into four bands: blue, orange, yellow, and red. The manufacturer promised that the human eye would blend the four colors and create the illusion of color. It didn't work. It made whatever was in the orange panel look orange and whatever was in the red section look red, and so on.

BATTER BITER

Baseball players have long worn mouth guards. For hitters, especially, a rubber or soft plastic oral appliance can provide a little extra protection and reassurance, with a 90-mile-per-hour fastball zipping dangerously close to one's face. But never before has a mouth guard actually helped to improve a batter's swing. Never, that is, before 1999, when a patent was issued for the Batter Biter.

The Batter Biter may look like an oversized infant's pacifier, dangling from a short cord that clips onto a ballplayer's jersey. But when properly used, it not only protects the batter's teeth, it also prevents him from turning his head, forcing him to keep his eye on the ball. By enforcing correct head positioning, the Batter Biter subtly reinforces proper body mechanics, giving hitters an extra edge.

The funny thing is, it might actually work. Baseball players are notoriously superstitious about hitting streaks. The great Wade Boggs ate a chicken dinner before every game to produce hits, while slugger Jason Giambi would don gold thong underwear to break a slump. The Batter Biter might conceivably give a similar psychological benefit, although it is awfully silly looking (albeit not any less so than a gold thong).

DOG WATCH

The idea of "dog years" is a simple ratio: Dogs live to a maximum of about 16 years, and humans live a maximum of 100 years or so, so this shorthand equates one calendar or "human" year with seven "dog years." It's not quite precise, but it helps all of us feel like our dogs live longer than they actually do.

But what about the dogs? Surely they are just as concerned about their mortality as we are about their mortality. With that in mind, in 1991 inventors Rodney Metts and Barry Thomas patented an alternative clock they called a "clock for keeping time at a rate other than human time." In short, the watch takes the concept of one year equaling seven dog years and extends that all the way down to seconds, minutes, and hours. It presents time in a way your dog can understand…if your dog can tell time, that is.

DEODORANT CANDY

Have you ever woken up in the morning and felt so lazy that you had absolutely no desire to cover up your naturally occurring, deeply offensive body stench, but still had just enough energy to eat some candy? Sure, we all have!

Thanks to Deo Perfume Candy, however, it's now possible to do the latter and still achieve the effects of the former. Available in regular and sugar-free varieties, Deo—which tastes like tangerine but contains a heaping helping of rose oil—utilizes the same biological principle which causes you to stink like a brewery the morning after downing a 12-pack of your favorite beer, but in a far more fragrant fashion, literally making your perspiration smell like flowers. It is probably worth noting, however, that Deo also necessarily contains geraniol, one of the primary chemicals in rose oil, which has a disconcerting tendency to attract bees. So, unless you have a profound desire to get stung, you may find it in your best interest to bypass the deodorant candy in favor of just getting your lazy butt out of bed and taking a shower.

BETTER LIVING THROUGH LAWN MOWING

In 2002 the German garden equipment maker Wolf-Garten introduced the Zero, a mower that cuts grass with lasers instead of blades. It's outfitted with a computer-guided array of four powerful lasers capable of cutting grass to an accuracy of 1 mm A stream of air then dumps the zapped blades back onto the lawn, so as to become super-futuristic mulch. The mower comes complete with a leather seat and WiFi, and costs $30,000.

But the future of grass-cutting isn't all high-tech. While some scientists prefer the space age power of lasers, others prefer the agrarian competence of farm animals. Scientists at Australian National University have figured out a way to turn the grass-hungry power of rabbits into a viable mowing product. Introducing the Rolling Rabbit Run, the world's first lawn mower powered entirely by rabbits. Constructed from bicycle wheels, chicken wire, and buckets, the device is basically a cylindrical rabbit cage that rolls around on the lawn as the rabbits eat the grass and fertilize it "naturally" (with their poop). Perfecting the mower took a little longer than expected because scientists couldn't get the rabbits to roll the cage on their own, because they'd used one male and one female rabbit, who kept giving up the grass to do the thing that rabbits famously do a lot. (In the new-and-improved version, two male rabbits are used.)

ROBO SKATER

We know you've been dying to break out your favorite pair of roller skates and boogie on down the sidewalk. Unfortunately, it's not the 1970s anymore, so put them back in the closet next to your platform shoes with aquariums in them. If only there were some way skating could be updated and made cool again, right?

Well, at least one inventor gave it a shot. Their solution? Full-body armor and lots of extra wheels. Like, 26 wheels in all. We're speaking of Robo Skater, which isn't nearly as neat as it sounds. Rather than some kind of robotic servant powered by roller skates, Robo Skater looks like the brainchild of someone who watched way too many episodes of *American Gladiators*.

In addition to the expected wheels on the bottom of the boots, Robo Skater's body armor is outfitted with wheels on the toes, elbows, bottom, back, chest, and hands. It seems likely that this product was designed for some bizarre sport competition that never materialized. It seems even more likely that the poor souls who test-marketed Robo Skater found the extra wheels helpful for when passing gangs of teenagers took the opportunity to push them down the nearest hill.

RON POPEIL'S GREATEST HITS

GLH Formula Number 9 Hair System. Popeil's baldness cure is really just temporary spray paint. ("GLH" stands for "great-looking hair.")

Inside-the-Shell Egg Scrambler. Popeil came up with the idea for this gizmo because he'd always been revolted when he would get a plate of scrambled eggs that weren't totally mixed, with the egg and white ribbons running concurrent. So he invented this motorized needle that fits inside of an egg and whips everything together, so you can crack it into the pan and make scrambled eggs the right way.

Mr. Microphone. You might remember the ad for this from 1970s television: A carload of guys passes a lovely lady listening to a radio, through which the driver broadcasts a message from his handheld microphone: "Hey, good lookin'! We'll be back to pick ya up later!" Mr. Microphone allowed a user to broadcast their voice over any FM radio–it was simply a very-short-range radio transmitter that would take over an FM signal... sort of, with lots of interference.

Smokeless Ashtray. A tiny battery-powered vacuum would suck in the smoke given off by a cigarette placed in the ashtray, then trap the smoke in a charcoal filter hidden inside.

AUGMENTED REALITY

Virtual reality? Oh, please. That is *so* passé. The preferred terminology to those in the know now is *augmented* reality, and the folks at Canon are doing their darndest to take the concept out of sci-fi films and bring it into the business world. What they're offering isn't a dream world, just a heightened version of the real one.

To experience Canon's MREAL system (that's short for "mixed reality"), you need only slip on the headset, which features two cameras, one in front of each eye, so that it's able to capture precisely what you'd be seeing if you weren't wearing the headset. Having done so, it then creates a 3D image which it combines with its own computer-generated material, resulting in an image which is equal parts real and fake but is hopefully accepted more as the former than the latter. Canon describes the device as producing "clear, solid-looking images with low distortion, reducing the impact of optical aberrations, even in peripheral areas, and further enhances the realism of the experience." Setting aside the technical gobbledygook, it's clear that Augmented Reality, while currently intended to help consumers imagine how, say, a piece of furniture might look in their home, is destined to end up being used for far more sordid scenarios. Either way, it's not likely to happen until the pricetag comes down considerably from its current $100,000 point.

MORE DUMB USB GADGETS

Gokiraji Cockroach: Want to see people scream at the sight of a cockroach without having to worry about bug-borne diseases? This fake insect moves across any surface via a USB–powered remote control. It also has a glow-in-the-dark backside.

Light Therapy: This device, which mounts onto a computer, simulates the light of a sunny day, aiding in the body's production of antidepressant hormones. It also comes with a car adapter, so that only the traffic can bum you out.

Yanko AromaUSB: Are you getting complaints at work about the smell of the microwaved leftovers you had for lunch? Simply fill the AromaUSB with any fragrance oil and plug it into your computer. A cool mist will emanate and give you the best-smelling cubicle at the office.

Thanko Necktie Clip Cooler: Outdoor formal events in the middle of the summer no longer mean profuse sweating in a suit. This tie clip includes a small fan that blows cool air upward, which means that men can keep cool and be fashionably cool at the same time, or as much as they can while awkwardly holding a laptop in front of themselves at an outdoor formal event.

iShaver: It's roughly the size of an iPhone, but flip the lid and it's an electric razor that gets recharged by plugging it into a USB port. (Just don't text with it–the last person who tried lost two fingers.)

THE DOUBLE CIGARETTE HOLDER

Have you ever seen an old movie where a glamorous socialite or elegant European lady takes a sultry drag on a cigarette held away and aloft of her face by a long, shiny cigarette holder? Those were a pretty fancy way to make smoking look extra-cool (smoking was cool before we knew it was deadly). But if everybody in the old-timey days, in this case the 1920s and 1930s, used cigarette holders, a truly fancy person had to set themselves apart from the hoi polloi with the next generation of cigarette holder: the *double* cigarette holder.

The idea began in books. Bulldog Drummond, the hero detective of a series of 1920s novels by Herman Cyril McNeile, carried around a cigarette holder that simultaneously burned and brought the smoke down. Bulldog's "held Turkish one side, Virginian the other." Because with the double cigarette holder, you didn't have to choose which kind of tobacco you wanted.

It looked like a pitchfork with two prongs.

ROOFIN' UP VERMONT

Winters are harsh in Winooski, Vermont; temperatures in the tiny town can drop as low as -20° F. As you can probably imagine, the heating bills around there tend to be a wee bit higher than the national average.

In 1979, during America's second major energy crisis in less than a decade, Winooski's residents grew increasingly frustrated with the rising cost of heating oil. The town's civic leaders decided to do something about it. During a dinner one night with his fellow bureaucrats, Mark Tigan, director of the Winooski Community Development Corporation, came up with a solution: Build a gigantic dome over the town.

Now keep in mind that this was the '70s. Sci-fi movies and novels of the time featured futuristic communities protected by domes. At the time, Tigan's idea didn't seem completely nuts. Still, nobody had ever really tried to do something like that before, so Tigan and Winooski would be real scientific pioneers.

Tigan had his staff quickly put together a study. They estimated that a one-square-mile dome over the town could reduce residents' heating bills by up to 90 percent annually. Still, Winooski's city council remained skeptical until Tigan and his staff pointed out that the project could attract a windfall in grant money from the U.S. Department of Housing and Urban Development. The people of Winooski were on board from the very

beginning, convinced that the dome would turn their frigid town into a tropical paradise, albeit an artificial one.

Tigan spent most of early 1980 coming up with solutions for all of the possible logistical problems. Vehicles with combustion engines would be banned from the dome, meaning residents would have to get around inside via electric cars or a yet-to-be-built monorail. Temperature would be regulated by pumping in air and heating it, or cooling it, with large fans. Tigan contacted a conceptual architect who drew up blueprints based on domed homes from around the world.

While officials at HUD wanted to fund the dome, President Jimmy Carter, who was facing a tough reelection campaign at the time and was worried that goofy projects like this might cost him a second term, personally nixed Tigan's plans. A citywide dome has yet to be built in the United States. (And Carter still lost.)

LINT LIZARD

Sometimes an invention ingeniously solves a problem and takes care of a need that humans have struggled with forever. Sometimes it's a modern improvement on an old design that makes life a little easier. And sometimes it's just a smaller version of an existing device that works just fine. Or it replaces an incredibly easy, minor chore that most people have no problem taking care of by hand.

The Lint Lizard falls into the last two categories. It's a tiny vacuum hose attachment you use to suck the lint out of a clothes dryer. Most lint accumulates in the lint trap, which you empty by hand after every load so as not to start a fire. Maybe you even occasionally vacuum it out, to get all the nooks and crannies in there clean, with your regular vacuum attachment. Well, stop it, because the Lint Lizard is here to replace that thing that worked just fine until now. But it's not merely for emptying the lint trap—you still have to do that. To use the Lint Lizard, you have to completely remove the lint trap and clean out the place where the lint trap rests.

But the real problem is that, according to online customer reviews, the Lint Lizard doesn't even work at doing the thing it's supposed to do—it doesn't suck very well, and it's only 30 inches long. And, since it attaches to a regular vacuum cleaner, you've still got to lug the vacuum cleaner into the laundry room and put it right up against the dryer.

FAKE EGG YOLK SLICER

Every single invention in this book is weird, but it takes a certain special something to transcend the simply unusual and soar into the realm of the truly esoteric. Let's honor Eustathios Vassiliou, the inventor of the device that automates the creation of artificial egg yolks, then cuts them into disc-shaped slices.

Is there really a demand for this gizmo? Are the people who eat artificial eggs really that picky about things like shape, appearance, or flavor? Do words like "extrusion-head" and "disk-cutting process" belong anywhere near our food chain? These are all excellent questions, and we don't really have answers for any of them, but we can tell you this: The minute anyone is in desperate need of a tube capable of squirting out mass quantities of artificial egg yolk and slicing it into disc-shaped segments, Vassiliou's legacy (he died in 2006) will be there to answer the call of duty. (For what it's worth, Vassiliou invented many things related to the creation and consumption of fake—but edible—eggish products.)

Until that day comes, we'll just have to keep on getting by with boring old regular eggs that were extruded the traditional way, with nary a cutting wire in sight. Actually, you know what? We might just be good with a bowl of cereal and some toast, thanks.

PSYCHOTRONIC WISHING MACHINE

We've detailed a lot of weird and bizarre gadgets and gizmos in this book, both the oddly useful and the completely useless. But whatever the ultimate result, somebody out there put in the time, money, and work to create something out of thin air.

They completely wasted their time. They should have just bought a Psychotronic Wishing Machine from Life Technology Research International. They could have used it to wish their invention into existence, or skip the middle man and just wish for fame and wealth.

So how does the Wishing Machine work? It doesn't. Okay, but here's how it operates: You simply speak into the microphone on the Machine to tell it what you want. Then sit back and wait a few days for your wish to come true. There's one big caveat from LTRI, though, to ensure wishes come true, and it's not what you think—make sure the machine is on, as wishes are less likely to come true if the machine is turned off while a wish is being processed.

Nevertheless, results are not guaranteed from the magic box that grants wishes via "conscious human interaction and energy fields" and also costs $499.

UNCLE JOHN'S BATHROOM READER CLASSIC SERIES

THE LAST PAGE

FELLOW BATHROOM READERS:

The fight for good bathroom reading should never be taken loosely—we must do our duty and sit firmly for what we believe in, even while the rest of the world is taking potshots at us.

We'll be brief. Now that we've proven we're not simply a flush-in-the-pan, we invite you to take the plunge:

Sit Down and Be Counted! Log on to www.bathroomreader.com and earn a permanent spot on the BRI honor roll!

If you like reading our books...
VISIT THE BRI'S WEBSITE!
www.bathroomreader.com

- Visit "The Throne Room"—a great place to read!
- Receive our irregular newsletters via e-mail
- Order additional *Bathroom Readers*
- Face us on Facebook
- Tweet us on Twitter
- Blog us on our blog

Go with the Flow...

Well, we're out of space, and when you've gotta go, you've gotta go. Tanks for all your support. Hope to hear from you soon.

Meanwhile, remember...

KEEP ON FLUSHIN'!